To Connie

INTRODUCTION

The use of materials has served as a measure of the progress of civilization: the Stone Age, the Bronze Age, the Iron Age. To the proponents of synthetic materials, this is the Age of Plastics. The complexity of modern civilization is so great that we are in the age of metals, the age of plastics, and the age of other types of materials as well, because all types of materials have found areas of usefulness in our life and work.

Because of the abundant flow of information about a wide variety of materials, information relevant to a particular group or type of materials tends to be scattered among various publications, and among different issues of the same publication. Compilations of articles on specific types of materials are thus a valuable literature resource, and the Materials Technology series of books embodies this concept. This series consists of compilations of articles pertinent to specific types of materials, drawn from journals published by Technomic Publishing Company.

Composite materials seek to utilize the most desirable features of widely differing types of materials. Reinforced composites are based on the use of high strength and high stiffness materials. Aluminum, steel, and organic reinforced epoxy systems represent important classes of composite materials.

The nine articles in this volume are reprinted from the Journal of Composite Materials, with the hope that this volume will prove to be a valuable literature resource in this field of knowledge.

CONTENTS

PUBLISHER'S NOTE

Effect of Particle Size on Dynamic Properties of Aluminum Powder – Epoxy Composites

R. SIERAKOWSKI, G. E. NEVILL, JR., AND D. HOWSE
Department of Engineering Science and Mechanics,
University of Florida, Gainesville, Florida 32601

(Received November 3, 1970)

Particle filled plastics have attracted interest for varied applications such as structural members and disposable injection molds. Recently, quartz particle-filled epoxy systems have been studied by Bodner [1, 2], for a single particle size and weight concentration of filler particles. The dynamic response of aluminum particle-filled epoxy composites, including the influence of particle size and volume percent of filler on material properties, is reported here.

SPECIMEN PREPARATION

Aluminum powder (Alcan Metal powder grade M144) was separated into the three bounded particle sizes shown in Table 1.

The matrix material was prepared using an epoxy mix of 10:1 by weight of Shell Epon Resin 815 to Curing Agent number 400. To insure uniform particle dispersion in the resulting mixture, particularly for the large particle sizes and low volume concentrations, the molds were inserted onto a rotisserie mechanism located inside the furnace chamber and rotated at approximately two RPM, during cure. The oven temperature was raised to the required curing temperature of 200°F prior to insertion and the specimens were cured for 90 minutes at this temperature. Only specimens which were essentially free of voids due to air bubbles were accepted. In addition, the density of the fabricated specimens was measured and only specimens which deviated less than .003 lbs/in^3 from the mean value for their group (same particle size and percent concentration) were used. Homogeneity of the specimens was examined visually, using a light microscope, at several longitudinal stations. Typical particle distributions for a fixed constituent percent and varying filler size are shown in Figure 1a-1c.

EXPERIMENTAL PROCEDURE

Composite specimens and specimens of pure epoxy used for reference were of length ½″ and diameter ½″, ±.003 inch. Compression tests were run at strain rates ranging from 10^{-5}/sec to 10^3/sec. The low strain rate tests utilized a Tinius Olsen testing machine while the higher rate tests were conducted using a Split Hopkinson Pressure Bar. Care was taken to minimize friction effects by lubrication

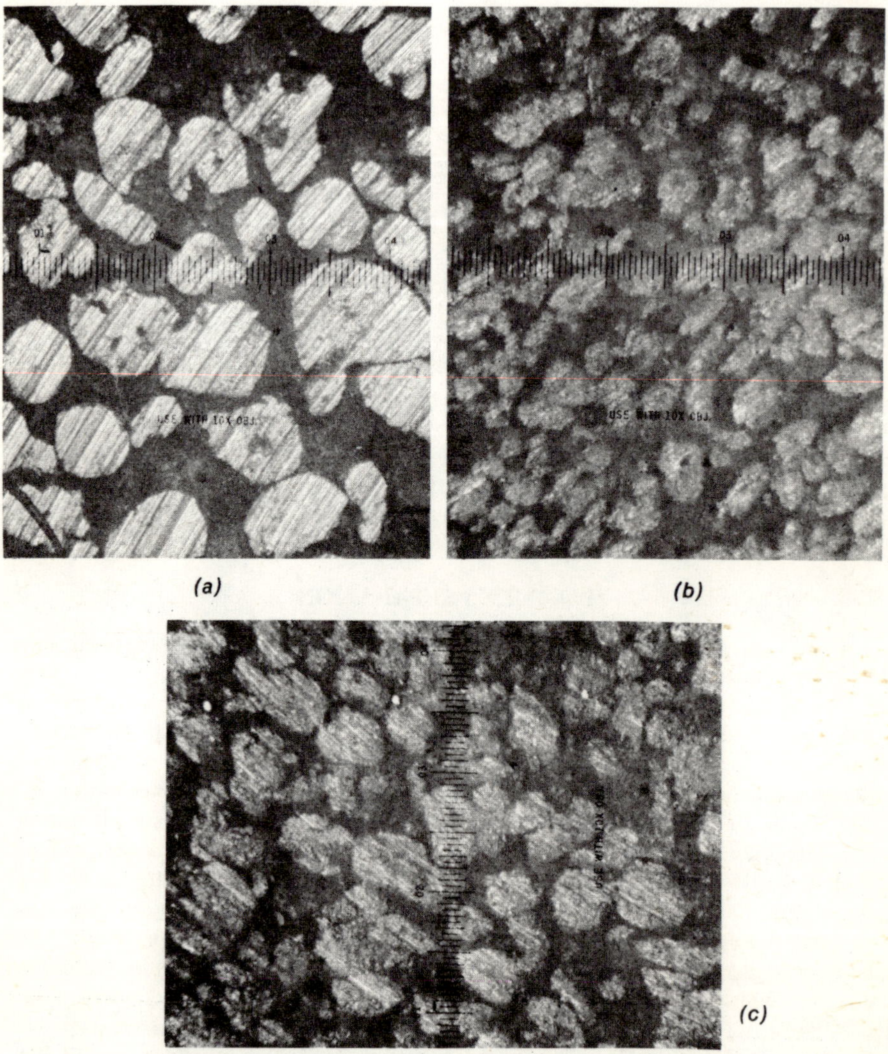

(a) (b) (c)

Figure 1. *Typical particle size distributions for a fixed percent of 36% by volume and 65% by weight of aluminum: (a) #60-80 particle size; (b) #100-140 particle size; (c) #140-200 particle size.*

Table 1

Particle Identification	U. S. Standard Sieve Series	Particle Size
1	#60-#80	.0098"-.0069"
2	#100-#140	.0069"-.0041"
3	#140-#200	.0041"-.0029"

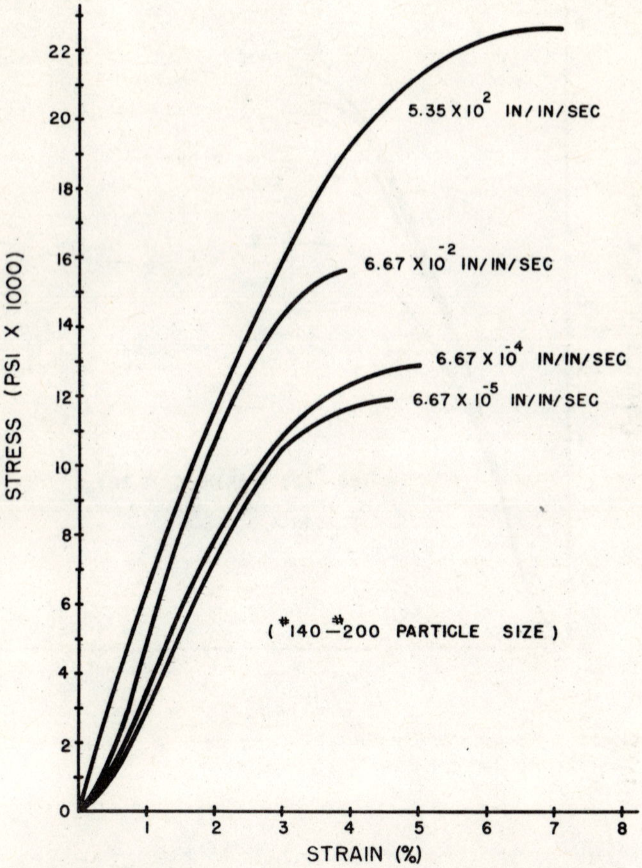

***Figure 2.** Stress-strain curves for fixed particle size and percent filler, 50% by weight (30% by volume) with varying strain rate.*

of the end surfaces of the specimens with a molybdenum disulfide grease. Deformation of the specimens was measured using a Tinius Olsen Type D-1 Deflectometer in the low rate tests. Operation of the Pressure Bar system is described by Lindholm [3].

DISCUSSION AND RESULTS

Data shown represents the average of three tests run for each particle size and percent of filler. Since the influence of several parameters, particle size, percent of filler and strain rate, is considered, each factor as it affects material response is discussed separately below.

In Figure 2 the influence of strain-rate on strength properties for a single

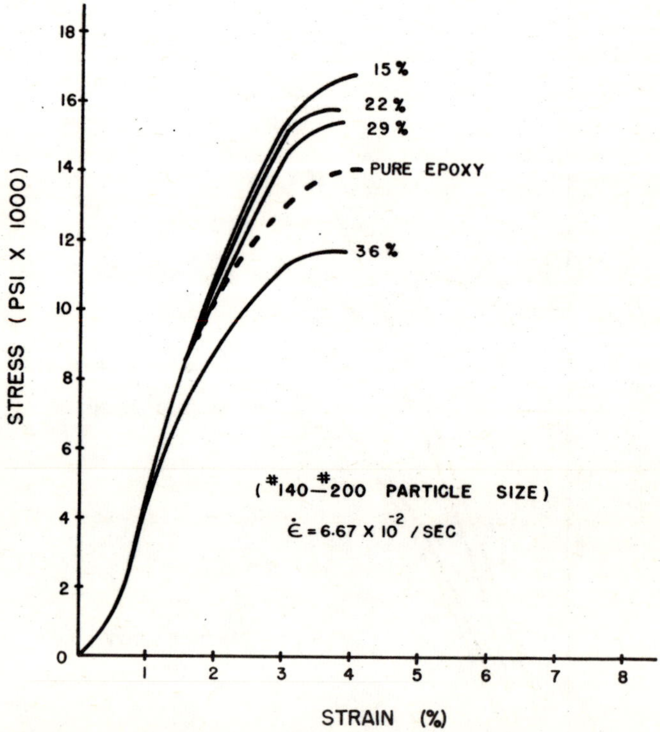

Figure 3. *Stress-strain curves for fixed particle size and strain rate, varying percent filler by volume.*

percent and particle size is shown. The strength is observed to increase by almost 100% between the lowest and highest strain rates. Very similar behavior was observed for the other percent filler and particle sizes tested. Figure 3 indicates the influence of percent filler on strength properties for a given strain-rate and particle size. The strength is observed to increase initially above that of the matrix material and then decrease as the percent of filler material is increased. The same trend was observed for the other particle sizes in the low strain rate region. A summary of this initial strengthening effect and the loss of reinforcement for a fixed particle size is shown in Figure 4. It is observed that for percents of filler material up to the order of fifty, material reinforcement is apparent at the low strain rates. However, at the higher strain rates, the influence of any reinforcing effect for the various percents tested appears negligible.

Since the trends are not completely consistent, the validity of results regarding particle size are questionable. Nevertheless it appears from results such as Figure 5, that there is an optimum particle size for maximum reinforcement effect in this range, namely the 100-140 mesh size.

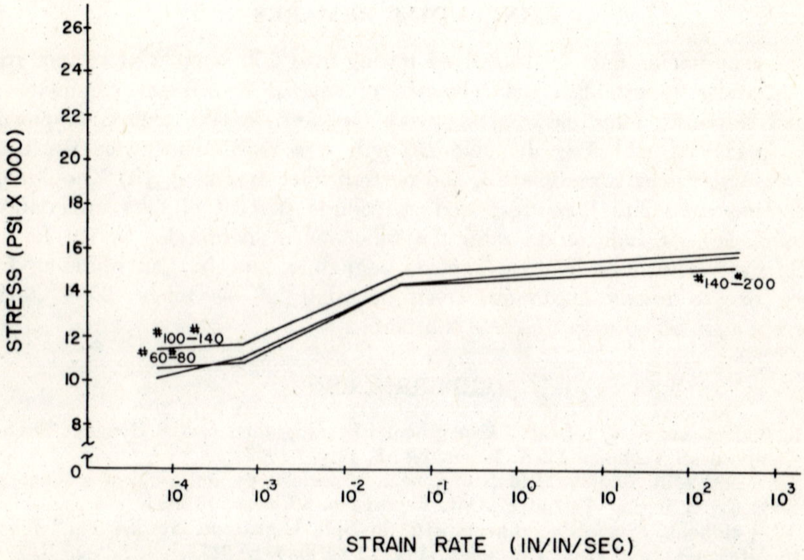

Figure 4. *Stress vs. strain rate for fixed percent (36% by volume) filler and varying particle size.*

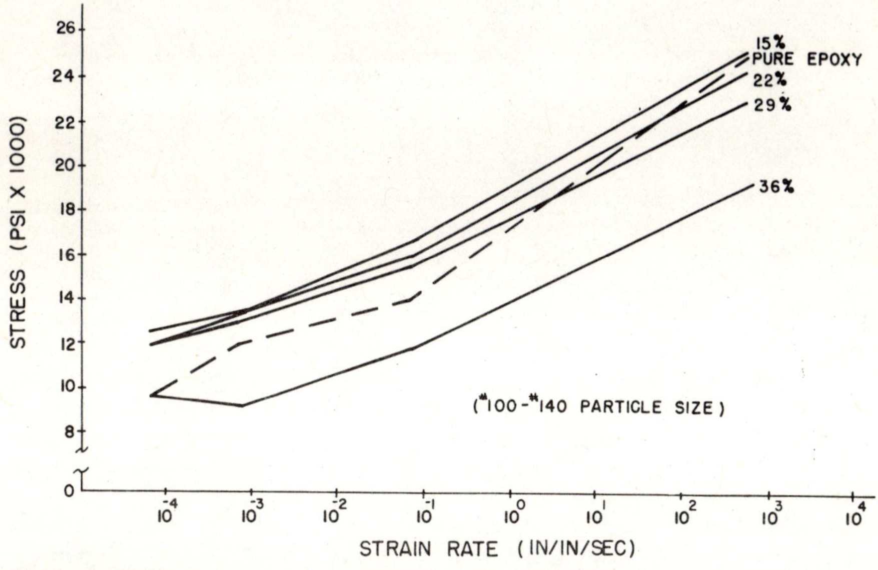

Figure 5. *Maximum stress vs. strain rate for fixed particle size, varying percent filler by volume and strain rate.*

CONCLUDING REMARKS

The experiments described involved testing over 150 samples at various strain rates in order to establish the influence of several constituent parameters on material response. Four basic trends were observed for the aluminum powder-epoxy specimens, (a) The dynamic strength was more than twice the quasi-static strength for each particle size and percent filler examined, (b) The strength initially increased and then decreased as volume percent of filler material was increased, (c) At high strain rates the influence of reinforcing action for any particle size and volume percent appears negligible, and (d) An optimum filler particle size to obtain maximum stress, bounded by the range .0069″-.0041″ appears to exist, all other factors held constant.

REFERENCES

1. S. R. Bodner and J. M. Lifshitz, "Experimental Investigations on the Dynamic Strength of Composites," Technion MML Report No. 8, July, (1967).
2. O. Ishai and S. R. Bodner, "Limits of Linear Viscoelasticity and Yield of a Filled and Unfilled Epoxy Resin," Technion MML Report No. 16, May, (1969).
3. U. W. Lindholm, "Some Experiments with the Split Hopkinson Pressure Bar," *Journal of the Mechanics and Physics of Solids,* Vol. 12 (1964), p. 317.

Dynamic Compressive Strength and Failure of Steel Reinforced Epoxy Composites

R. L. Sierakowski, G. E. Nevill, Jr.,
C. A. Ross and E. R. Jones

*University of Florida
Gainesville, Florida*

(Received February 12, 1971)

ABSTRACT

Results of an experimental program systematically evaluating the deformation and fracture of steel wire reinforced epoxy composite systems are presented. The program involved mechanical testing in the strain rate range 10^{-5} to 10^3/sec and impact testing of composite specimens using massive elastic targets at strain rates approximately 10^4/sec. Specific results presented include static and dynamic properties, strain rate sensitivity, information on the nature and character of dynamic fracture, influence of specimen geometry and reinforcement spacing, and high speed photographs of dynamic failure modes. Further, a simplified energy criterion is proposed for predicting failure modes and critical velocities.

INTRODUCTION

Many factors influence the performance of fiber reinforced systems. These included constituent physical and mechanical properties, interfacial bonding, and geometrical properties including size, shape and orientation of reinforcements. In addition, crack initiation and propagation behavior through continuous and discontinuous reinforced composites are important.

The basic features of material failure under static loading may be entirely different than under dynamic loading for monolithic materials. For composite materials, characterization and response are even more complicated. In addition, the type of load, tensile, compressive, shear or bending, becomes an important consideration. In this investigation, principal attention was focused on static and dynamic compressive loading of composite specimens. Some work dealing with static loading problems is reported in [1-5]. For compressive loading of composite systems, several modes of failure have been observed including elastic buckling, local buckling and crippling, crack extension in the direction of the loaded filament, and separation of individual fibers from the surrounding matrix. The latter failure modes can be associated with crack propagation; however, whether the compressive loading causes local buckling preceding crack propagation, or whether crack initiation and propagation contribute to the buckling phenomenon is not well understood.

Recently, investigations into dynamic properties of composite materials, both analytical and experimental, are receiving increased attention. Since the dynamic properties evaluation, and wave propagation phenomena are interrelated, in that understanding one requires knowledge of the other, attention to both is important for understanding and evaluating material behavior.

It is apparent that under dynamic loading complex failure phenomena may be present. A general survey of the dynamic properties and wave propagation problem for materials, with sections pertaining to composites, is contained in [6]. Some representative analytical studies of the wave propagation problem are reported for materials characterized as anisotropic in nature in [7-9], and for materials with point dependent properties in [10-14]. Changes in wave characteristics during propagation in fiber reinforced composites have been treated to some extent in [15-16].

The use of composite materials for dynamic applications is increasing despite the limited quantity of experimental information on their dynamic properties and failure or fracture behavior. This investigation was conducted to obtain such information. The reinforcing elements were selected in order to produce representative and reproducible continuous fiber reinforced matrix systems. Geometrical variations among specimens included fiber diameter, fiber spacing and volume percent of reinforcement. Included in the results is a qualitative description of the failure/fracture phenomena based on visual observation of the impact of composite specimens striking a rigid elastic target. The experimental program involved low strain rate compression testing (in the order of 10^{-5} to 10^{-1}/sec), using a Tinius Olsen Testing Machine and high strain rate testing (in the order of 10^{3}/sec), using a Split Hopkinson Pressure Bar system.

The impact testing utilized an air gun. High speed photographs using an image-converting camera provided visual observation of the dynamic failure and fracture characteristics of the specimens tested. Results presented include qualitative information on the nature and character of dynamic failure plus quantitative evaluation of the importance of specimen geometry, reinforcement spacing and density, composite component property relations and resultant energy to failure.

SPECIMEN FABRICATION

It was necessary to develop a capability for fabricating the required composite systems since samples with the desired variations in size of reinforcing elements, volume percent reinforcements, spacing and geometry, were not commercially available.

The specimens were fabricated using a winding machine, shown schematically in Figure 1, designed for this program. The machine is used to wind composite blocks from which were cut an average of 20 cylindrical specimens with nominal dimensions of 0.375 inch diameter by one-half inch in length.

Fully annealed type 304 stainless steel wires were wound on a flat mandrel assembly having nominal dimensions of 0.375 inch by 1.625 inches by 2.50 inches with a thickness of 0.25 inch for the end plate assemblies. A threaded shaft was used to align the filaments on the mandrel; the filaments were guided onto the shaft assembly by a pulley system carried by the shaft proper. The gears could be changed to provide various filament spacings and the wires were pretensioned to insure accurate and uniform placement.

Layers were wound on the mandrel to a depth of approximately 7/16 inch. The mandrel was then removed and cleaned in a trichlorethyene vapor degreaser and placed in an oven at 150°F for one-half hour to remove traces of cleaning solvent. The specimens were then impregnated with epoxy resin, Shell Epon 828, heated to a temperature of 150°F, and mixed with curing agent D in the ratio of 12 parts by weight of curing agent to 100 parts of epoxy resin. To minimize voids within the specimen, the mandrel assembly was also heated to 150°F, and was slowly immersed in the resin mixture in a vacuum chamber having a pressure of about one inch Hg absolute.

After impregnation and curing of the specimens at 150°F for 2 hours, the composite blocks were cut and removed from the mandrel block. Specimens were then machined to the dimensions 0.375 inch in diameter by 0.50 inch in length. Two mandrels were wound for each desired size and volume fraction with approximately 30 specimens obtained from each assembly. Typical fabricated specimens for a particular volume fraction or reinforcement are shown in Figure 2, and a summary of the various filament sizes and volume fractions tested is included in Table 1. The accuracy of filament spacing my be observed from Figure 2.

Fabrication of the cross-piled specimens was accomplished in the same manner as above with modification of the mandrel to allow for 90° rotation during the winding process.

Table 1. Volume Fractions and Wire Sizes Tested.

Volume Percent, $\overline{V}_f$	***10***	***26***	***40***
Wire Size	0.004 0.008 0.016	0.004 0.008 0.016	0.004 0.008 0.016

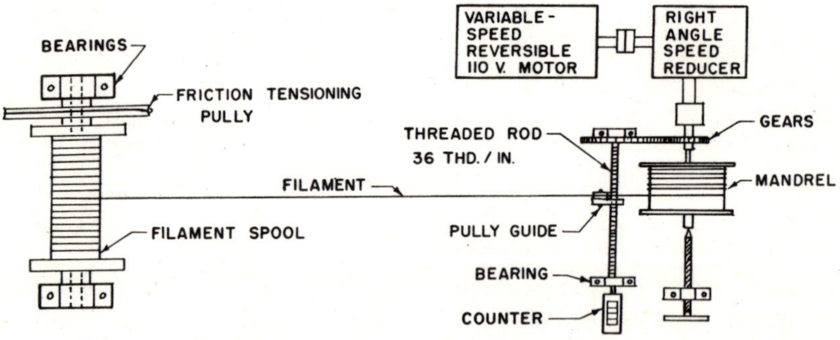

Figure 1. *Schematic of winding apparatus.*

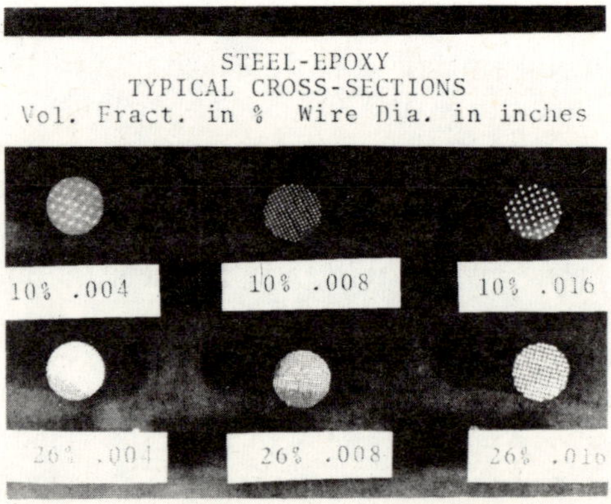

Figure 2. *Typical composite test specimens.*

EXPERIMENTAL TECHNIQUES

The prepared specimens were tested in uniaxial compression both statically and dynamically under varying load and strain rates at room temperature approximately 24°C. Thus the epoxy matrix was in the glassy elastic state. Specimen dimensions were accurate to ±0.001 inch and end surfaces were ground flat and parallel. During testing, the ends were lubricated with molybdenum disulfide grease to minimize friction effects.

Material characteristics at the low strain rates, 10^{-5} to 10^{-10}/sec, were measured using a Tinius Olsen Testing Machine. Load was measured by a bonded strain gage load cell, and strain measurements were based on relative crosshead movement as determined by an Olsen Type D-1 Deflectometer.

In order to examine composite response at strain rates of approximately 10^3/sec, a Split Hopkinson Pressure Bar assembly was used. This device

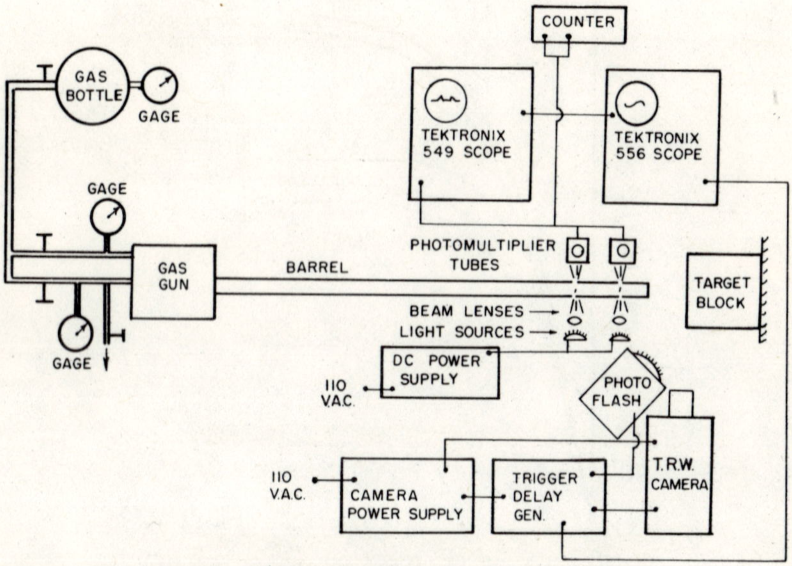

Figure 3. *Overall schematic of gas gun assembly.*

utilizes measurements of elastic waves in hardened steel pressure bars to determine the relative motion of the two faces of the test specimens and the associated stress. A full description of its utilization for obtaining dynamic properties data is included in reference [17].

To study further the failure of the steel reinforced epoxy system, an air gun assembly, modified from that described in [18], was constructed. This assembly consists of the following components: the air gun control panel and related equipment, a velocity measuring system, and a high speed photographic system for recording the impact event. The complete assembly is shown schematically in Figure 3, and a description of the air gun assembly and its development is included in [19]. Velocity was determined from measurements of the time taken for the specimen to cut parallel light beams a fixed distance apart. The output of the photocells was recorded on an oscilloscope; an independent check on specimen velocity was made using a digital counter, which gave the measured time in microseconds.

Photographs of the impact event were obtained with a TRW sub-microsecond framing camera. This camera is an image converting type, which can be used to take a sequence of three or five frames, with variable exposure times in the nanosecond range and interframe delay times in the microsecond range.

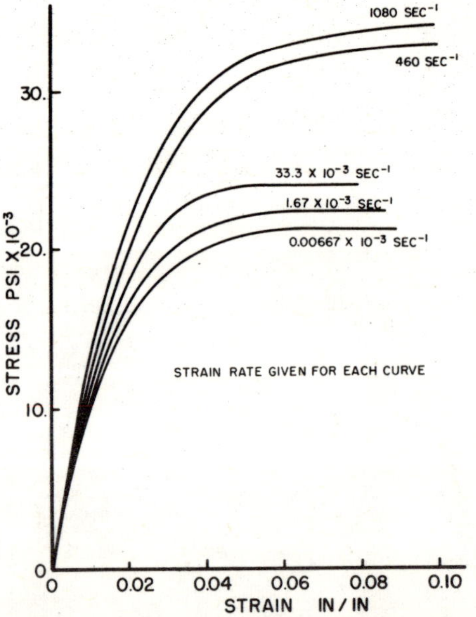

***Figure 4.** Stress-strain curves for steel-epoxy composites, $\overline{V}_f = 10\%$ wire diameter = 0.004 inch.*

RESULTS AND DISCUSSION

Static Compressive Properties:

A series of low strain rate tests [6.7×10^{-5}/sec] was performed to examine the nature of the resulting stress-strain relations, the maximum stress associated with specimen failure, the modes of failure in compression considering fiber orientation, and the maximum stress associated with wire size and volume percent reinforcement.

The reinforcement volume fractions and wire sizes selected have been give in Table 1. A typical sequence of stress-strain relations obtained for a single volume percent reinforcement with varying wire size is shown in Figures 4-6. The curves shown represent the average results of several samples (normally three) tested at each strain rate. Figure 7 displays the maximum specimen stress developed as influenced by wire size and volume percent reinforcement for specimens loaded parallel to the reinforcing direction. The results show that at low strain rates, for a fixed volume percent, high maximum stress values are associated with smaller wire sizes, while increasing the volume percent for a given wire size gives higher maximum stress values. Similar results have been observed for intermediate strain rates in the order of 10^{-2}/sec.

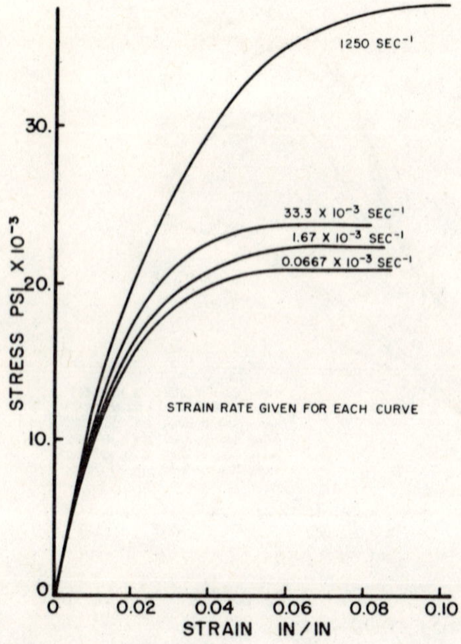

Figure 5. *Stress-strain curves for steel-epoxy composites, $\overline{V}_f = 10\%$ wire diameter = 0.008 inch.*

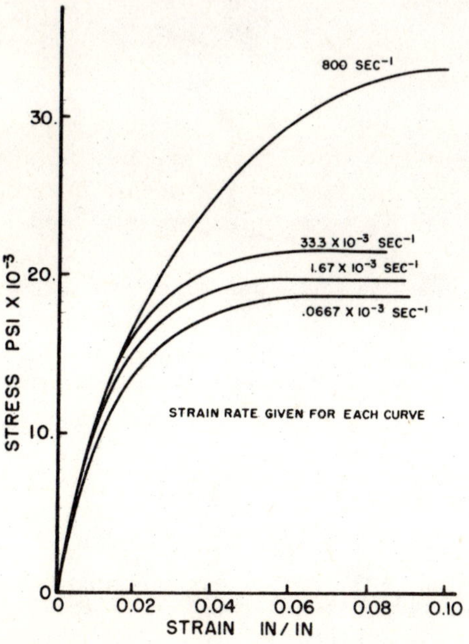

Figure 6. *Stress-strain curves for steel-epoxy composites, $\overline{V}_f = 10\%$ wire diameter = 0.016 inch.*

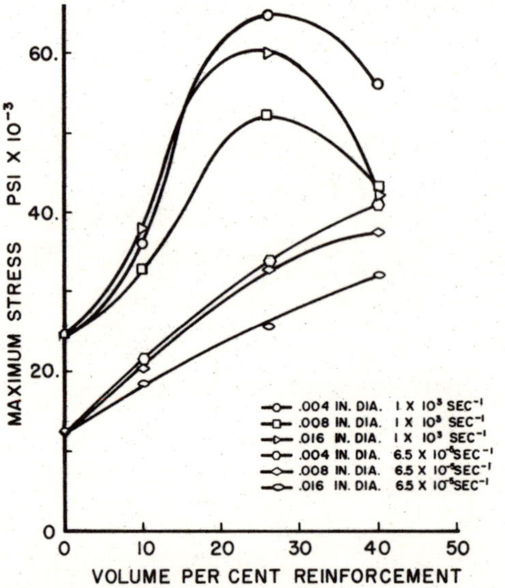

Figure 7. *Ultimate compressive strength of steel-epoxy composites.*

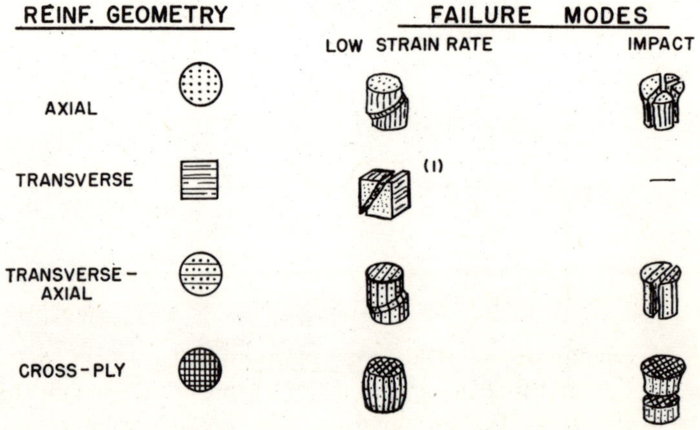

Figure 8. *Compressive failure modes for composite materials.*

Discussions of possible analytical models for predicting compressive failure stresses together with experimental results are included in references [20-24]. Many of the proposed models are based on planar filament buckling considering the filament to be a beam on an elastic foundation. This is a high energy deformation mode which yields compressive stress predictions far greater than experimental results [19, 23, 24]. Figure 8 schematically displays several of the final static compressive failure modes obtained in this

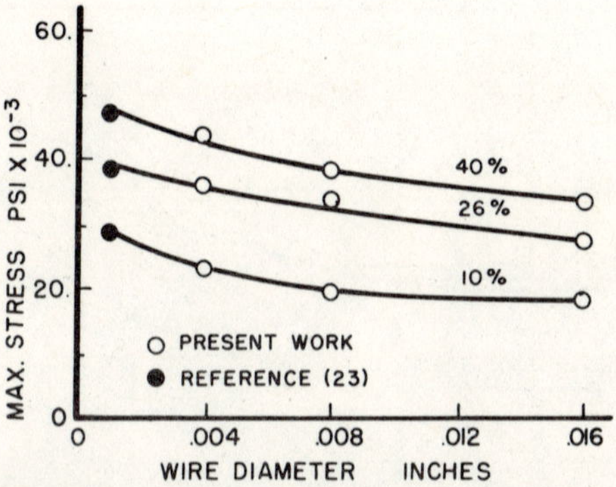

***Figure 9.** Ultimate compressive strength vs. wire diameter for steel-epoxy composites.*

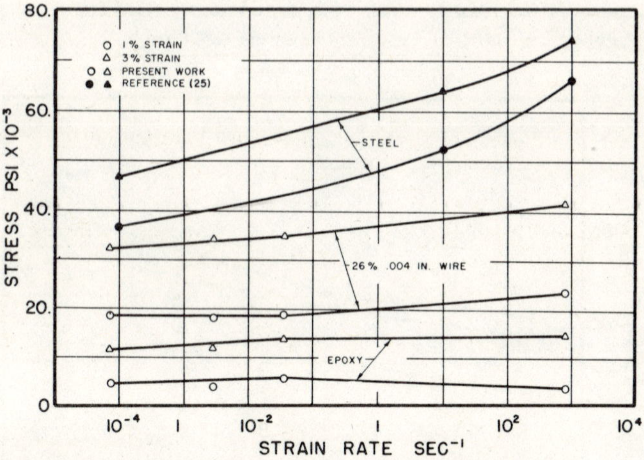

***Figure 10.** Composite strain rate dependence, steel-epoxy composites, $\overline{V}_f = 26\%$ wire diameter $= 0.004$ inch.*

program for various types of fiber orientations with respect to the loading directions. It is noted that the addition of the transverse filaments to the axial filaments clearly weaken the composite. For the low volume percent reinforced composites loaded in a direction parallel to the fiber orientation, the final failure mode was presaged by specimen barrelling. Also included are results obtained for Glass-Epoxy specimens tested in compression with filaments oriented in the transverse direction [24]. The results reported here coincide closely with those presented in [23] where studies on both ductile and brittle types of reinforcements with unordered filament packing are described; similarity in the observed compressive failure characteristics is noted particularly. This is further demonstrated in Figure 9 where current results are presented together with data from [23] at comparable volume percents

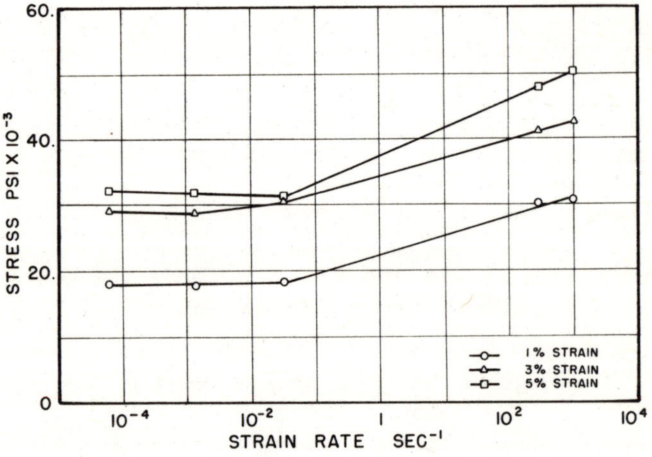

Figure 11. *Composite strain rate dependence, steel-epoxy composites, $\overline{V}_f = 26\%$ wire diameter $= 0.008$ inch.*

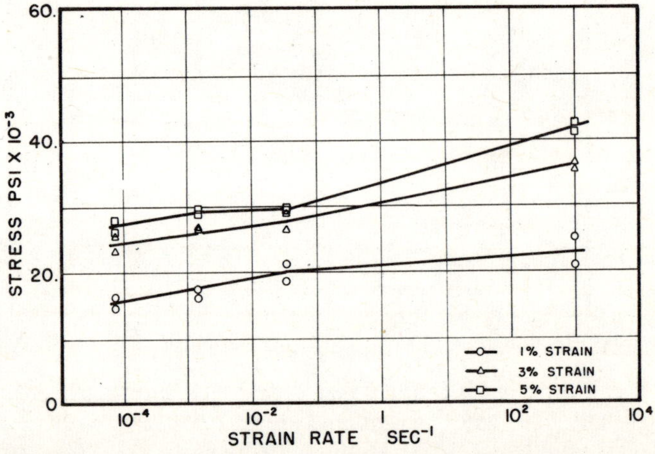

Figure 12. *Composite strain rate dependence, steel-epoxy composites, $\overline{V}_f = 26\%$ wire diameter $= 0.016$ inch.*

and rates of testing. The increased strength associated with decreasing filament size is noted. This may be attributable to the greater amount of surface bonding area between filament and matrix and thus more effective bonding.

Dynamic Compressive Properties:

A typical sequence of dynamic stress-strain curves are also shown in Figures 4-6. The composite dynamic strength exceeds its static strength by as much as one hundred percent.

In Figures 10-12 the influence of strain rate is presented for composites with

$\overline{V}_f = 26\%$; results for other values of $\overline{V}_f$ were similar. For comparison purposes, the base matrix material properties and those of steel as reported in reference [25] are also shown. The latter reports on the same type wire as is used in the present test series. It is observed that for a given volume percent reinforcement, the stress increases with strain rate, the influence of the latter becoming more significant in the high strain rate region.

In Figure 7 the maximum stress at failure is shown for data obtained from the Split Hopkinson Pressure Bar tests. Note that a maximum failure stress occurs from each wire size at the intermediate volume percent. Also, for higher volume fractions of reinforcing material less strain rate sensitivity is apparent. This is possibly due to the reduced interfiber spacing resulting in greater matrix shear stresses and thus easier matrix fracture.

The failure modes observed for strain rates up to 10^{-1}/sec followed closely the quasi-static data. Failure modes in tests with the Hopkinson Bar (strain rates 10^{3}/sec] were unobtainable because of the test procedure which resulted in complete specimen fracture. Documentation of failure modes at strain rate 10^{4}/sec was obtained from impact tests which are discussed in the following section.

Failure Dynamic Properties:

A series of experiments involving the impact of flat ended cylindrical specimens against essentially rigid targets was also performed. The test equipment used has been discussed briefly in the previous section, Experimental Techniques, and more thoroughly in reference [19]. Critical specimen velocity and qualitative fracture behavior of composites based on visual records were the principal failure characteristics studied.

In contrast to conventional monolithic materials, the compressive failure behavior of composites can be varied and complex. For the steel-epoxy systems, two distinct types or stages of failure were identified. The first stage was associated with filament delamination/debonding, characterized by the onset of permanent bending or buckling of individual filaments. This occurs at that point of the stress-strain curve for which the composite maximum stress is first attained. This occurrence has been correlated with the horizontal tangent to the stress-strain curve. The second distinct stage of failure observed was complete fracture and fragmentation of the specimen. For this case, the first observable fragmentation was considered sufficient to define failure.

An effort was also made to predict critical velocities in impact tests using Hopkinson Pressure Bar data. Several theoretical approaches were examined for their suitability for making such predictions. Based upon similarity in the experimentally determined response characteristics ($\sigma - \epsilon$) for the dynamically compression tested composites with tensile tested monolithic materials [26], a critical velocity in compression was observed. From the data obtained, a simple energy criterion was established for calculating the critical

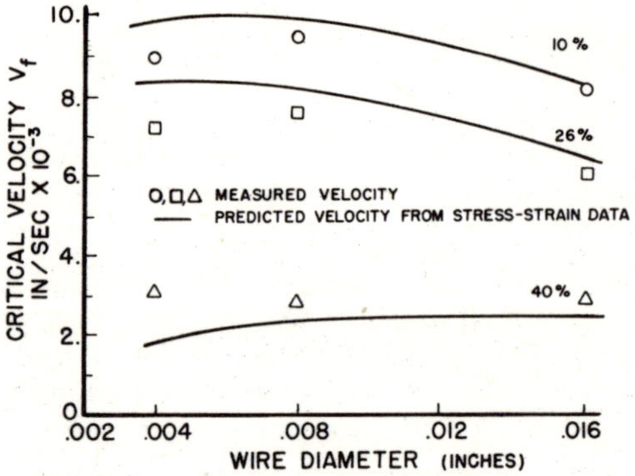

Figure 13. *Critical fracture velocity vs. wire diameter for steel-epoxy composites.*

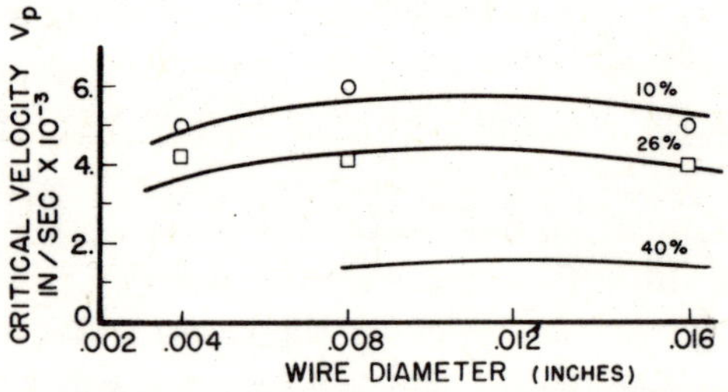

Figure 14. *Critical plastic flow velocity vs. wire diameter for steel-epoxy composites.*

velocity by correlating analytically the dynamic pressure bar data with the gas gun experiments. The analysis was based on equating the impact kinetic energy of the specimen to the strain energy as obtained from stress-strain data using the Hopkinson Pressure Bar. A tabulation of this data is presented in Table 2 and a separate graphical summary of theoretically predicted critical velocities with experimentally determined critical velocities is shown in Figures 13-14. The solid curves represent the calculated prediction, while actual test results are shown as data points for the various specimens tested. It is noted that two critical velocities are shown V_P representing the critical velocity for initiation of failure stage one, as previously discussed, and V_F the velocity associated with failure stage two.

Data obtained provided good correlation between predicted and actual test results for initial specimen delamination and ultimate specimen fracture. The experimental points shown represent interpolated data for impact velocities since precise critical velocity determination was not possible. It is noted that experimental data for the critical velocity V_P for the forty volume percent specimens is absent. This is due to the increased embrittlement of the steel epoxy specimens with increase in volume percent of reinforcing filament such that only specimen fracture could be observed. This effect is confirmed by the mechanical properties data obtained. In the results for both V_P and V_F there appears to be an optimum wire size for a particular volume percent of reinforcement which will generally yield an optimum critical velocity.

It should be noted that the strain rates in the gas gun experiments were approximately one order of magnitude greater than the corresponding Hopkinson Pressure Bar tests (the strain rates using the air gun were approximately 10^4 in/in/sec). However, the area under the curve is sufficiently insensitive to such a strain rate change that the pressure bar results were used without strain rate correction.

For all the specimens which were tested on the Hopkinson Bar and by the air gun assembly, a high speed photographic series of the specimen failure and projectile impact, plus post-test photos were taken. These photo sequences were then used to determine the dynamic failure mechanisms and to compare these mechanisms with those observed in the quasi-static and high strain rate pressure bar tests.

Typical photographs of the dynamic failure and fracture behavior of these uniaxial composites are shown in Figure 15. The failure modes observed in the gun tests appeared to exhibit several features distinct from those observed in the low and intermediate strain rate tests. In particular, for the low volume fraction specimens, for the various wire sizes tested and for low speed impact, a localized deformation was observed. However, as the impact velocity was increased fragmentation of the specimens occurred. For the transverse axial

Table 2. Critical Velocities for Steel-Epoxy Composite Projectiles.

Specimen Dimensions: Diameter 0.050"
Length 1.00"

Composite		V_P(in/sec)		V_F(in/sec)	
Volume %	Wire Diameter	calc.	meas.	calc.	meas.
10	0.004"	4500	5000	9880	9000**
10	0.008"	5350	6000*	8960	9050**
10	0.016"	5120	5000	8280	8250
26	0.004"	3460	4200	7760	7200
26	0.008"	4350	3800	8100	7000
26	0.016"	3890	4000	6300	6000**
45	0.008"	1420	***	2370	2830
45	0.016"	1300	***	2600	2980

* Lowest velocity tested.
** Highest velocity tested.
*** Measurable plastic flow does not occur.

Figure 15. *Dynamic fracture of steel-epoxy composite with $\overline{V}_f = 26\%$, wire diameter = 0.016 inch. Impact speed is 8000 in/sec.*

and cross ply specimens tested dynamic failure was associated with specimen splitting and brittle-like separation, respectively. The various failure modes are shown schematically in Figure 8.

CONCLUDING REMARKS

The dynamic compressive properties of uniaxial, transverse axial and cross ply steel reinforced epoxy specimens have been investigated at varying strain rates for continuous reinforced composite systems. The principal results obtained are included in the following paragraphs.

The influence of strain rate on composite material response has been recorded. For the composites tested, rate effects have been found to depend upon composite constituent properties, geometric arrangement, volume fraction of reinforcement and size of reinforcing fibers.

The compressive failure modes for low strain rate testing are varied and distinct from tensile failure modes. Further, the high strain rate compressive failure modes are significantly different than those at the lower rates.

The maximum failure stress has been found to depend upon wire size, volume percent of filament reinforcement and strain rate. As the rate of loading is increased, the maximum failure stress is increased. For low speed tests the maximum stress increases with volume percent reinforcement and the smaller wire sizes have greater load capacity. For the high strain rate tests an optimum volume percent reinforcement exists for the various wire sizes tested; however, the smaller wire size again yields the greatest strength. Composite specimens with reinforcements other than in the load direction displayed maximum failure stress at both low and high loading rates approximately twenty percent below that of the uniaxial specimens.

For composite specimens with a brittle matrix, a simple energy criterion has been found meaningful for predicting specimen separation and delamination. The high strain rate or impact failure/fracture characteristics are often distinctly different from observed failure/fracture in low strain rate testing.

Since composites can be quite rate sensitive, dynamic as well as static test results are essential to predicting impact processes. Mechanical properties obtained at varying strain rates are thus necessary if design is to be meaningful where wave propagation effects are significant.

ACKNOWLEDGMENTS

The authors wish to acknowledge the support of the Air Force Armament Laboratories, Elgin Air Force Base through contract F08635-68-C-0115 and to express appreciation for the advice and encouragement provided during this research effort.

REFERENCES

1. Society of Aerospace Material and Process Engineers, *Symposium on Filament Winding*, Vol. 1, Pasadena, California, March 28-30, 1961.
2. Symposium, *Testing Techniques for Filament Reinforced Plastics*, Air Force Materials Laboratory, Wright-Patterson Air Force Base, Dayton, Ohio, September 21-23, 1966.
3. Society of Aerospace Material and Process Engineers, *Symposium on Advanced Fibrous Reinforced Composites*, Vol. 10, San Diego, California, November 9-11, 1966.
4. L. Holliday, *Composite Materials*, Elsevier Publishing Company, London (1966).
5. L. J. Broutman and R. H. Krock (eds.), *Modern Composite Materials*, Addison Wesley Publishing Company, Reading, Massachusetts (1967).
6. T. Nicolas, *The Mechanics of Ballistic Impact, A Survey*, Air Force Materials Laboratory, Technical Report 67-208, Wright-Patterson Air Force Base, Dayton, Ohio, July, 1967.
7. W. M. Ewing, W. S. Jardetzky, and F. Press, *Elastic Waves in Layered Media*, McGraw-Hill Book Co. (1957).
8. M. J. P. Musgrave, "Elastic Waves in Anisotropic Media," *Progress in Solid Mechanics*, Edited by I. N. Sneddon and R. Hill, Vol. 2, North Holland Publishing Company,

Amsterdam (1961), pp. 61-85.

9. P. C. Chou, "Introduction to Wave Propagation in Composite Materials," *Composite Materials Workshop,* Edited by S. Tsai, J. C. Halpin, and N. J. Pagand, Technomic Publishing Company, Stamford, Connecticut (1968), pp. 193-216.
10. U. S. Lindholm and K. D. Doshi, "Wave Propagation in an Elastic Nonhomogeneous Bar of Finite Length," *Journal of Applied Mechanics,* Vol. 32 (1965), p. 135.
11. R. G. Payton, "Elastic Wave Propagation in an Elastic Rod," *Quarterly Journal of Mechanics and Applied Mathematics,* Vol. 19 (1966), p. 83.
12. G. Eason, "Wave Propagation in Inhomogeneous Elastic Media," *Bulletin of the Seismological Society of America,* Vol. 57 (1967), p. 1267.
13. D. P. Reddy, "Stress Waves in Nonhomogeneous Elastic Rods," *Journal of the Accoustical Society,* Vol. 34 (1969), p. 1273.
14. C. T. Sun, K. C. Valanis and R. L. Sierakowski, "The Extension of a Superposition Principle to Wave Propagation in Nonhomogeneous Elastic Media," *Journal of Composite Materials,* Vol. 3 (1969), p. 454.
15. J. D. Achenbach and G. Hermann, "Dispersion of Free Harmonic Waves in Fiber-Reinforced Composites," *Journal of the American Institute of Aeronautics and Astronautics,* Vol. 6 (1968), p. 1832.
16. H. D. McNiven and Y. Mergi, "Controlled Dispersion of Axisymmetric Waves in Composite Rods," *Journal of the Acoustical Society,* Vol. 43 (1968), p. 691.
17. U. S. Lindholm, "Some Experiments with the Split Hopkinson Pressure Bar," *Journal of the Mechanics and Physics of Solids,* Vol. 12 (1964), p. 317.
18. R. Plunkett, D. E. Godfrey, K. Yamaishi and P. Beynet, "A Gas Gun for Transonic Velocities," Air Force Materials Laboratory, Technical Report AFML-TR-69-345, Wright-Patterson Air Force Base, Dayton, Ohio, March, 1970.
19. R. L. Sierakowski, G. E. Nevill, Jr., C. A. Ross and E. R. Jones, "Studies on the Ballistic Impact of Composite Materials," Air Force Armament Laboratory, Technical Report 69-99, July, 1969.
20. B. W. Rosen, "Mechanics of Composite Strengthening," Space Sciences Laboratory, General Electric Company, Presented at American Society of Materials Seminar on Fiber Composite Materials, Philadelphia, Pennsylvania, October 17, 1964.
21. R. L. Foye, "Compressive Strength of Unidirectional Composites," Paper presented at the American Institute of Aeronautics and Astronautics Meeting, January, 1966.
22. N. Fried, "The Response of Orthogonal Filament Wound Materials to Compressive Stress," *Proceedings, 20th Conference, Society of Plastic Industry,* Reinforced Plastics Division (1965).
23. E. Moncunill de Ferran and B. Harris, "Compression Strength of Polyester Resin Reinforced with Steel Wires," *Journal of Composite Materials,* Vol. 4 (1970), p. 62.
24. O. Ishai, A. E. Moehlenpah and A. Preis, "Yield and Failure of Glass-Epoxy Composites," *Journal of the Engineering Mechanics Division,* ASCE, Vol. 96 (1970), p. 739.
25. U. S. Lindholm and R. L. Bessey, "A Survey of Rate Dependent Strength Properties of Metals," AFML-TR-69-119, April (1969).
26. J. S. Rinehart and J. Pearson, *Behavior of Metals under Impulsive Loads,* American Society of Metals, Cleveland, Ohio (1954).

Some Damping Properties of Aluminum Particle Filled Epoxy Composites

R. Sierakowski, G. Hemp, and A. Hokstad, *Department of Engineering Science and Mechanics, University of Florida, Gainesville, Florida 32601*

(Received February 21, 1971)

Some recent damping and modulus results obtained for quartz particle filled composites, have been described by Bodner and Lifshitz [1]. The damping data, obtained over a limited frequency range, was found to be little effected by filler content. Recent, additional studies investigating the damping of glass filament reinforced epoxy composites have been reported by Schultz and Tsai [2] considering frequency and amplitude dependency as well as filament orientation. Further results for glass epoxy and boron epoxy specimens have been discussed by Tauchert and Moon [3]. In the present investigation, the results of some introductory tests considering the influence of particle size and percent filler on the damping properties and elastic modulus are presented for aluminum powder-epoxy composites.

MATERIALS AND TEST METHODS

Material specimens with an epoxy matrix and an aluminum particle filler were prepared using the three particle sizes indicated in Table I. The volume percentage of particle filler was varied using 13%, 30% and 40% (these corresponded to 25%, 50% and 60% filler by weight). The method of fabrication and the precautions taken to insure uniform distribution of particles throughout the specimens and to prevent voids are discussed in [4]. All specimens were ¼ inch square in cross-section and averaged 8½ inches in length.

The specimens were dynamically tested using a Brüel and Kjaer complex modulus apparatus, model 3930, to determine the damping properties (reported in terms of the loss factor d). Associated equipment required for the tests included an oscillator (B&K 1022), an amplifier (B&K 2603) and a level recorder (B&K 2305). A schematic of this apparatus as used in the present experiments is shown in Figure 1.

Table 1.

U. S. Standard Sieve Series	Particle Size
#60-#80	0.0098"-0.0069"
#100-#140	0.0069"-0.0041"
#140-#200	0.0041"-0.0029"

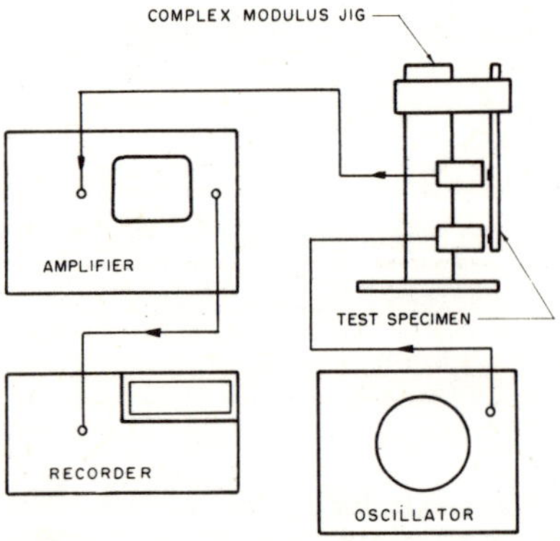

***Figure 1.** Test equipment schematic.*

Since the materials were non-magnetic, small ferro-magnetic discs were attached to the specimen near the tip and mid length of the cantilever beam. The specimens were excited using a magnetic transducer at the station near the tip and the response signal was monitored from the station near mid length. The tests were performed at the experimentally determined fundamental frequency of each specimen and care was taken to adjust the beam length to restrict the maximum root strain at resonance to the range 50-500 μin/in as reported in [2]. to the aluminum-alloy matrix.

Three methods of determining the material damping were used; these are: (1) the band width method, $d = \Delta f/f$ where Δf is the half-power band width of the resonance peak and f is the fundamental frequency; (2) the reverberation or log decay method, $d = -\dfrac{0.0366}{f}\dfrac{dy}{dt}$, where $\dfrac{dy}{dt}$ is the rate of decay of the response measured in db/sec; and (3) the log decrement, $d = \dfrac{1}{\pi N}\ln(x_1/x_N)$. Further discussion of these methods is available in [5].

Tests on each specimen were repeated a minimum of three times in order to report an average value for each method of determining the loss factor. The results showed a high degree of repeatability. In addition, the influence of elapsed time from specimen fabrication to test was considered; all tests were conducted within 70-75 hours of fabrication time. Longer time intervals were selected for comparison purposes; the recorded results showed little variation from the main tests indicating that if time elapsed is a factor it is within the first 75 hours.

After recording data for the loss factor, a series of nondestructive tensile tests to determine the material elastic modulus were performed at slow strain rates on a Tinius Olsen Testing Machine. This data was then evaluated for comparison with predicted upper and lower bound estimates for the elastic moduli of particle reinforced composites as discussed in [6]. In addition, the real part of the complex modulus was calculated using the measured natural frequencies in the equation $E = \dfrac{4\pi^2\mu f^2}{I\beta^4}$ where μ is the mass per unit length and β is the first eigenvalue of the frequency equation governing flexural vibrations. The use of this equation appears satisfactory for the case of small damping and low frequencies [3].

RESULTS AND DISCUSSION

Figures 2 and 3 show the test results for the influence of particle size and volume percent filler on the loss factor d. Results of each method for the determination

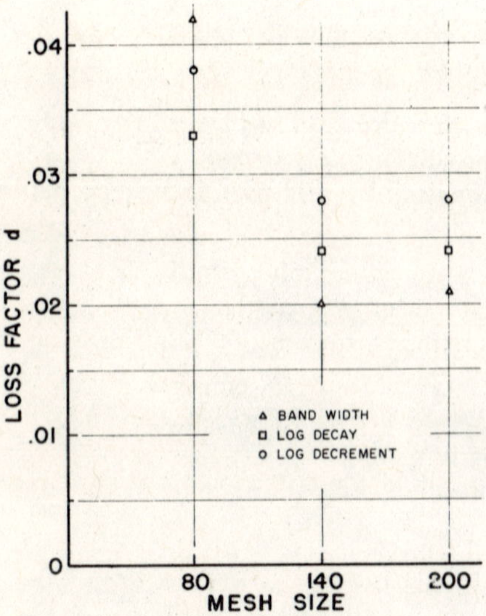

Figure 2. *Variation in loss factor with particle size; 40% volume filler.*

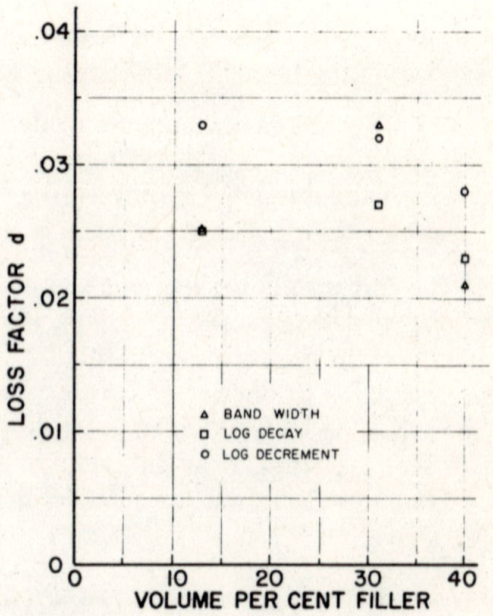

Figure 3. *Variation in loss factor with volume % filler; mesh size 200.*

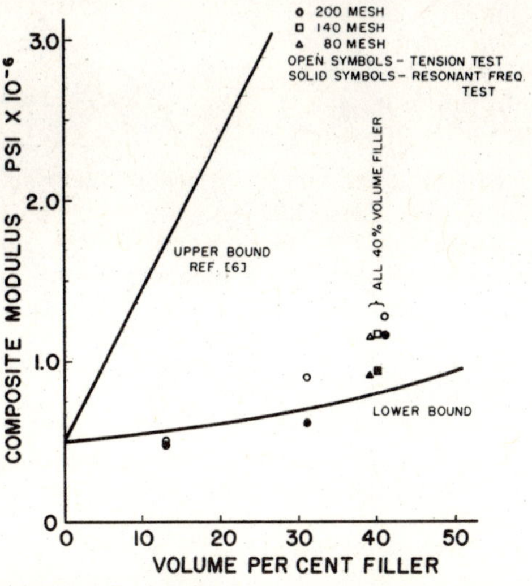

Figure 4. *Variation of composite modulus with particle size and volume percent filler.*

of d are shown. Although there may be more variation than desired in values obtained from the methods used, two observations can be made:

(1) The loss factor appears to decrease with a decrease in particle size while holding the volume percentage of filler constant in the specimen.
(2) The loss factor appears to be less influenced by a change in volume percentage than a change in particle size.

The test results for the material elastic modulus are shown in Figure 4 for the various particle sizes and percent of filler. Values of E calculated from the resonant vibration tests are consistently smaller than the results of the tensile tests. Theoretical curves for the upper and lower bounds are presented for reference. Figure 4 shows that the lower bound is a much better representation of the experimental data than the upper bound in the range of tests performed.

More extensive tests are planned to evaluate the validity of hypotheses (1) and (2) above.

ACKNOWLEDGMENTS

The authors wish to acknowledge the contributions of Mr. A. Gregalot for his repetition and refinement of the above test procedures while participating in an NSF Undergraduate Research Program during 1969.

REFERENCES

1. S. R. Bodner and J. M. Lifshitz, "Experimental Investigations on the Dynamic Strength of Composites," *Mechanics of Composite Materials*, ed. F. W. Wendt, H. Liebowitz and N. Perrone, Pergamon (1970), p. 663.
2. A. B. Schultz and S. W. Tsai, "Dynamic Moduli and Damping Ratios in Fiber-Reinforced Composites," *Journal of Composite Materials*, Vol. 2 (1968), p. 368.
3. T. R. Tauchert and F. C. Moon, "Propagation of Stress Waves In Fiber-Reinforced Composite Rods," Proc. of the AIAA/ASME 11th Structures, Structural Dynamics, and Materials Conference (1970).
4. R. Sierakowski, G. E. Nevill, Jr., and D. Howse, "Effect of Particle Size on Dynamic Properties of Epoxy Composites," *Journal of Composite Materials*, Vol. 5 (1971), p. 118.
5. R. Plunkett, "Measurement of Damping," *Structural Damping*, J. E. Ruzicka Edition ASME Publication (1959), pp. 117-131.
6. G. P. Sendeckyj, "Statistical Theory of Elastic Moduli of Particulate Composites," *Journal of Composite Materials*, Vol. 3 (1969), p. 177.

Strain Rate Effect on the Ultimate Tensile Stress of Fiber Epoxy Strands*

T. T. CHIAO AND R. L. MOORE, *Lawrence Radiation Laboratory, University of California, Livermore, California 94550*

(Received October 24, 1970)

The strain rate effect on composite performance has been the subject of many arguments. This note points out some of the problems involved in measuring strain rate effect and presents our tensile data on some simple unidirectional composites.

EXPERIMENTAL

For this study we used only one spool of single end glass (SCG 150-1/0-1Z-HTS 901, 204 filaments). The spool was kept in a container with a drying agent when not in use. A flexible epoxy formulation in 40% acetone solution (Dow DER 332/Union Carbide ERL-4206/Celanese Epi-Cure 855,† 70 parts/30/40), was used in the program. This low viscosity resin, which has higher elongation to break than the fiber, and a torsional shear strength of 3.8 Ksi, can be cured at room temperature; however, in order to save time, we evaporated off the acetone and cured the specimens at 350°F for one hour. The properties of the cast pure resin will be determined in connection with our work on stress-rupture of composites.

A complete description of our specimen fabrication and test method for fiber strands was reported previously [1]. Briefly, a strand is cut from the spool of fiber, allowed to untwist by holding one end up, dipped in a disposable trough containing epoxy-acetone solution, and cured vertically under constant low tension in an oven. Through the use of clamps and an alignment fixture, the specimen is then fixed to a 10 in. gage length. The clamps (cushioned with the combination of pressure-sensitive aluminum foil, 0.015-in. Neoprene sheet, and household cement) are held by four 6-32 screws, each having 50 to 75 in.-oz. torque.

We tested the specimen-clamp system in an Instron tensile test machine, model TTDM, with a load capacity of 10,000 kg. At the higher crosshead speeds, i.e., 2-20 in./min, the conventional load-weighing system is completely inadequate, particularly for fibers of low failure strain.

---

* Work performed under the auspices of the U.S. Atomic Energy Commission.

† Reference to a company or a product name does not imply approval or recommendation of the product by the University of California or the U.S. Atomic Energy Commission to the exclusion of others that may be suitable.

Table 1. Strain Rate Effect on the Ultimate Tensile Stress of Fiber/Epoxy Strands.

Material	Strain Rate (in./in.-min)	Av. Tensile Strength (psi)	95% Conf. Limits	Std. Dev. (psi)	Coeff. of Var. (%)	No. of Specimens
S-Glass/Epoxy[a]	1.97×10^{-4}	497,700	±15,400	21,500	4.3	10
	1.97×10^{-3}	522,900	±18,000	25,000	4.8	10
	1.97×10^{-2}	555,100	±15,000	21,000	3.8	10
	1.97×10^{-1}	607,500	±11,400	15,900	2.6	10
	1.97	610,300	±11,800	16,500	2.7	10
Graphite/Epoxy[b]	1.97×10^{-4}	225,700	± 5,800	8,200	3.7	10
	1.97×10^{-3}	224,000	± 7,400	10,300	4.6	10
	1.97×10^{-2}	218,800	± 5,200	11,100	5.1	20
	1.97×10^{-1}	231,800	± 5,600	11,900	5.1	20
	1.97	237,100	± 5,800	10,500	4.4	15
Boron[c]	1.97×10^{-4}	469,000	± 8,100	11,000	2.4	10
	1.97×10^{-3}	466,800	±17,700	25,800	5.6	10
	1.97×10^{-2}	464,600	±33,900	47,200	10.3	10
	1.97×10^{-1}	466,800	±80,000	104,000	22.4	10
	1.97	469,000	±39,100	82,600	17.7	20
Beryllium	5×10^{-4}	155,700	± 400	600	0.3	10
	1.97×10^{-3}	159,000	± 600	800	0.6	10
Wire[d]	5×10^{-3}	160,600	± 800	1,200	0.8	10
	1.97×10^{-2}	163,900	± 500	700	0.4	10
	1.97×10^{-1}	166,300	± 700	800	0.6	10
	1.97	168,700	± 1,000	1,200	0.7	8
Steel Wire[e]	1.97×10^{-4}	529,200	± 3,600	3,100	0.5	5
	1.97×10^{-3}	527,100	± 300	200	--	5
	1.97×10^{-2}	527,100	± 300	200	--	5
	1.97×10^{-1}	527,100	± 4,700	3,600	0.7	5
	1.97	533,300	± 6,700	5,200	1.0	5

(a) Owens Corning Corp., Hts-901, cross sectioned area, $A = 2.14 \times 10^{-5}$ in.^{2}

(b) Hitco, treated HMG 50 fiber, $A = 6.39 \times 10^{-5}$ in.^{2}

(c) Texaco Experiment Inc., 4-mil Borofil, $A = 1.356 \times 10^{-3}$ in.^{2}

(d) The Beryllium Corp., 4-mil wire, $A = 1.233 \times 10^{-3}$ in.^{2}

(e) Crucible Steel Co., 5-mil wire, $A = 1.935 \times 10^{-3}$ in.^{2}

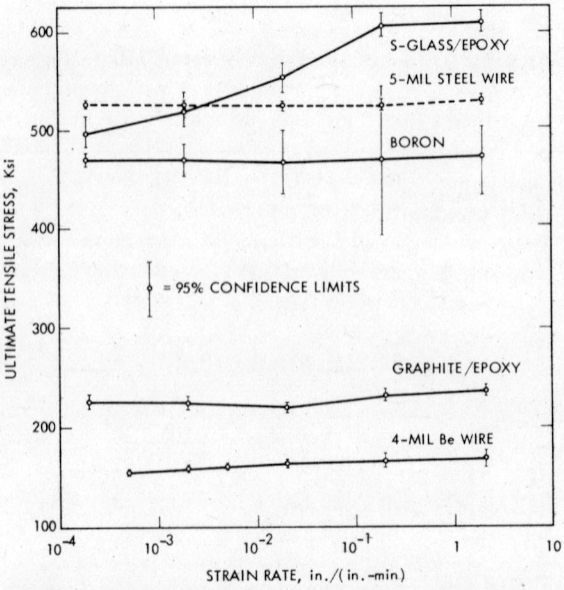

Figure 1. ***Strain rate effect on the ultimate tensile stress of fiber/epoxy strands.***

Instead of the commonly-used bridge amplifier-recorder system, we excited the load cell with a regulated D.C. power supply never exceeding 4 volts. The load cell output was amplified about 100 times in a D.C. amplifier and recorded on an oscillograph.

RESULTS AND DISCUSSION

Table 1 summarizes the effect of strain rate on the ultimate failure tensile stress of several common fiber/epoxy strands. The same data is plotted in Figure 1. What we call strain rate in this study is really the crosshead speed of the tensile machine divided by the gage length of the specimen, 10.0 ± 0.02 inches. This definition, of course, is subject to debate. However, considering the relatively low breaking loads of the strands vs. the machine capacity, and the impracticality of using strain gages on so many strands, we feel that it is justifiable for comparative purposes.

For mono-filament fibers, such as boron (B), beryllium (Be) and steel, we wish to point out that there was no epoxy coating on the specimens. Our prior work (Ref. 1) verified that matrix had no effect on single-filament B and Be fibers.

Figure 1 shows that the strain rate does influence the ultimate failure stress of the S-glass/epoxy strand. For all practical purposes, however, we can ignore the rate effect on the B, Be, steel, and graphite/epoxy strands, at least within the limited speed range of most standard tensile machines. Our curve on S-glass/epoxy strands shows the same trend, but with less scatter, than the results on E-glass/epoxy unidirectional laminates obtained by Ishai et al. [2]. It also agrees well with the very limited data of Pirzadeh and Kennedy on S-glass/epoxy strands [3]. On advanced fibers, our results do support Rauch's comments in his recent book [4]: "The cross-head speed used in testing filaments is not generally critical because most of the new, advanced fibers are not particularly strain-rate sensitive in the range normally used for testing (approximately 0.0014 to 0.1 min^{-1}). It is important, however, not to apply the load so rapidly that the response of the recorder becomes a limiting factor." However, he reported no data.

Further examining the S-glass/epoxy data from an engineering point of view, we find that a designer can ignore the strain rate effect and make no more than 10% error for each ten-fold change in rate.

SOME PROBLEMS

In measuring strain rate effect, we found that the speed of response of the standard recording instruments was too slow to indicate the true readings. This matter has too often been overlooked by many laboratories. Actually, we were forced to study the rate effect of composites ourselves after two independent contracting laboratories reported conflicting data—one showed that the ultimate tensile stress of S-glass/epoxy strands decreased with strain rate; the other indicated that there was no effect. We believe that they both had instrumentation problems.

Another problem in studying composites is that the fabrication of the test specimens often is not under control. This is particularly true of specimens of complex form, such as laminates or filament wound articles. We therefore selected the simple fiber/epoxy strand to avoid this problem. Hopefully, the simple specimen can approach a uniform state of stress during testing.

REFERENCES

1. T. T. Chiao and R. L. Moore, "A Tensile Test Method for Fibers," *J. Comp. Mat.*, Vol. 4 (1970), p. 118.
2. O. Ishai, A. E. Moehlenpah, and A. Preis, "Temperature and Time Effects on Yield and Failure of Unidirectional Glass Epoxy Composites," Monsanto/Washington University, ONR/ARPA Association, AD 861189 (July 1969).
3. N. Pirzadeh and P. B. Kennedy, *Symposium on Standards for Filament Wound Reinforced Plastics*, ASTM STP-327 (1962), p. 174.
4. H. W. Rauch, *Ceramic Fibers and Fibrous Composite Materials*, Academic Press, New York (1968), p. 34.

Steady Wave Analysis of Wave Propagation in Laminates and Mechanical Mixtures*

D. E. MUNSON AND K. W. SCHULER

Sandia Laboratories, Albuquerque, N. M. 87115

(Received April 3, 1971)

ABSTRACT

A mechanical theory is developed for the propagation of steady waves in laminated media. This analysis treats wave propagation parallel and perpendicular to the laminate plates. It predicts the response of the composite from a knowledge of the component volume fractions and mechanical equation of states. Under the restriction of fluid or hydrodynamic materials, the results yield a hydrodynamic prediction which is independent of composite geometry. Consequently, the analysis is equally applicable to mechanical mixtures. Theoretical calculations are compared to experimental Hugoniot results for a cloth laminate quartz phenolic, a plate laminate and three mechanical mixtures of Al_2O_3 in epoxy. Excellent agreement is obtained at low stresses, if the expected deviations caused by material strength are recognized. Thermodynamic effects caused by deviations of the loading path from that of a single shock become more predominant as the stress increases. Models describing both the strength and thermodynamic effects are discussed.

INTRODUCTION

THE STUDY OF WAVE PROPAGATION in composites can be conveniently divided into a low and high pressure regime. In the low pressure regime the shear strengths of the component and the bond strengths of interfaces are not exceeded and the behavior of the component materials can be considered linearly elastic. Thus, a complete description of wave propagation in this regime requires the solution of boundary value problems in linear elasticity theory whose complexity is directly related to the geometric complexity of the composite [1, 2].

In the high pressure regime the hydrostatic component of the applied stress field is usually many times greater than the shear strength of the components or the bond strength of the interfaces. Thus in this regime the

* This work was supported by the United States Atomic Energy Commission.

behavior of the component materials may be considered fluid or hydrodynamic and generally will be nonlinear. Even with this simplification to hydrodynamic behavior, a complete description of wave propagation requires that the geometric details of the composite be considered. Carrying out such a study in all but the simplest of geometries is prohibitive. There is, however, an important class of problems in which the geometric structure of the composite is "fine" and allows the reverberation of waves between components to occur in times which are short compared to the duration of the applied stress pulse. For this class of problem it seems that a steady state will be reached at large distances from the front of the pulse, and further that conditions in this steady state should be relatable by a steady wave analysis. Here we present a purely mechanical steady wave analysis which is applicable to waves propagating in laminated plates both normal and perpendicular to the laminations.

The original motivation for this work stemmed from the steady wave analysis of Tsou and Chou [3] which was restricted to waves propagating along the laminates of a laminated plate composite. In order to solve the steady wave problem, they found it necessary to include explicitly the energy balance relation for a steady wave and to make assumptions on the heat flow between components. Since the exact stress strain path of a particular component is difficult to define, precise treatment of heat flow among components may not add sufficient exactness to warrant the increased calculational problems. It appears, as we will later confirm, a steady wave analysis may be developed on a purely mechanical basis with the thermodynamics implied through the selection of the constitutive relationship.

A steady wave analysis is developed for hydrodynamic behavior of an *n*-component, repetitive laminate. This permits calculation of the constitutive relationship of the composite from knowledge of constitutive relationships of the components. The analysis for wave propagation normal to the laminates is found to be identical to that for propagation parallel to the laminates. Results of the analysis are compared to calculations of Tsou and Chou [3] and to experimentally determined Hugoniots of a cloth laminate quartz-phenolic and a particulate mixture of Al_2O_3-epoxy and to an experimentally obtained wave profile in a stainless steel-epoxy plate laminate. With the exception of deviations at low stress levels, which would be expected, comparison between experiment and theory is quite good. Models are suggested to account for the influence of material strength at low stresses and for the influence of loading path on component internal energies at high stresses.

THEORY

We consider two extremes of plane wave propagation in layered composites of large extent—in one case, the direction of propagation is along the

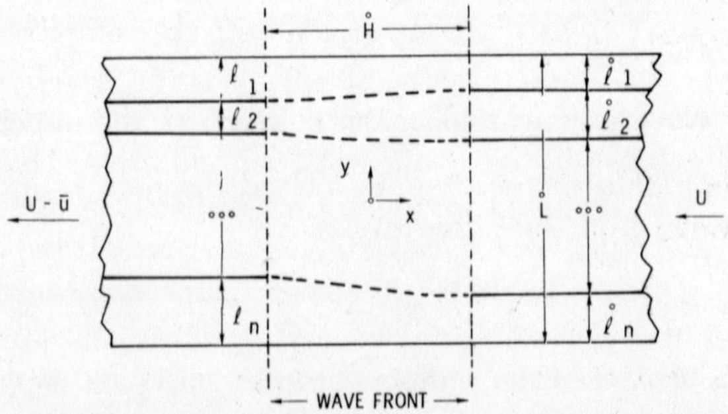

Figure 1. General laminate section, x-direction propagation illustrated.

layers; while in the other case, it is perpendicular to the layers. Figure 1 shows a typical portion of the composite in which the laminate sequence of n layers is repeated regularly. Initial layer thicknesses are $\overset{0}{l}_n$ and the repetition unit is $\overset{0}{L}$. Initial volume fractions of each layer are specified by $\overset{0}{\alpha}_n = \overset{0}{l}_n/\overset{0}{L}$. For propagation in the x-direction, it seems reasonable to assume that lateral reverberations in the plates will damp out and, in regions distant from the wave front, a steady state will exist. For propagation in the y-direction, it should be possible, as the thicknesses of the plates become small, to realize an average wave speed and to calculate averaged quantities behind the wave front. In both cases, we assume that the pressures in the steady state or equilibrium state will be the same in each component. Since these assumptions neglect all details of structure, and hence, the effects of geometric dispersion, we expect to obtain an analysis which is a first order approximation. For this reason, we shall take the simplest constitutive relationships for the mechanical behavior of the components which retains material nonlinearities. In particular, each component is assumed to behave hydrodynamically; i.e., supports no shear stress. Further, thermodynamics are neglected except for the requirement that the shock hydrostat [4] be taken as the equation of the state of the components. Thus during the compression process the temperature of the components will change due to shock heating. In general the temperature change will be different in each component resulting in a situation where heat flow could occur. Since we do not wish to treat thermodynamics explicitly, we assume a condition of no heat flow. This assumption seems reasonable in light of the fact that significant heat conduction occurs at times long compared to characteristic wave propagation times. The change in energy of the component also appears implicitly.

In the following we shall consider, in turn, steady wave analysis in both the x and y direction. Without loss of generality, we require the section to have unit depth in the z-direction and conclude from symmetry considerations that this dimension cannot change during wave propagation.

Steady Wave Analysis:

For steady wave propagation in the x-direction, the upper and lower surfaces of the repetition section remain a fixed distance $\overset{0}{L}$ apart. Initially, far ahead of the wave front (superscript zeros)

$$\Sigma \overset{0}{l}_n = \overset{0}{L} . \tag{1}$$

After lateral movement of the component interfaces have come to equilibrium behind the wave front we have

$$\Sigma l_n = \overset{0}{L} , \tag{2}$$

where l_n is the final thickness of the nth layer. The components take on a *new* density ρ_n. It follows directly from the definition of volume fraction that

$$\Sigma \overset{0}{\alpha}_n = \Sigma \alpha_n = 1 . \tag{3}$$

Because of the steady wave treatment, it becomes convenient to choose a coordinate system moving with the velocity, U, of the steady wave. In this system, the front is stationary and, far ahead of the front, material appears to flow into the front with velocity, U. Far behind the front, material flows away with velocity, $U - \bar{u}$, where $\bar{u}$ is the steady wave particle velocity. We assume $\bar{u}$ is the same in every layer of the composite.

Balance of the mass of each component, M_n, per unit time requires $\frac{\partial}{\partial t}(M_n) = 0$, which implies that the flow of any component is the same into and away from the steady wave front. This may be written for each component as

$$U \overset{0}{l}_n \overset{0}{\rho}_n = (U - \bar{u}) \, l_n \rho_n . \tag{4}$$

Dividing by ρ_n and summing over all components gives

$$U \overset{0}{L} \Sigma \frac{\overset{0}{\alpha}_n \overset{0}{\rho}_n}{\rho_n} = (U - \bar{u}) \overset{0}{L} ,$$

or

$$\bar{u} = U \left(1 - \Sigma \frac{\overset{0}{\alpha}_n \overset{0}{\rho}_n}{\rho_n} \right). \tag{5}$$

Balance of momentum for a steady wave is independent of time, which requires that $\frac{\partial P}{\partial X} + \frac{M_n}{l_n} \frac{\partial u}{\partial X} = 0$, where u is the particle velocity and P is the pressure. However, for a unit of time, $M_n = U\, l_n \rho_n$ and, as a result [5], $l_n \dfrac{\partial P}{\partial X} + M_n \dfrac{\partial u}{\partial X} = 0$. Adding and subtracting $P \dfrac{\partial l_n}{\partial x}$, we have

$$\frac{\partial}{\partial x}(l_n P + M_n u_n) - P \frac{\partial l_n}{\partial x} = 0. \tag{6}$$

Summing over all layers,

$$\frac{\partial}{\partial x}\Sigma(l_n P + M_n u_n) - P \frac{\partial \Sigma l_n}{\partial x} = 0,$$

but $\dfrac{\partial \Sigma l_n}{\partial x} = \dfrac{\partial \overset{0}{L}}{\partial x} \equiv 0$ and, since the wave is steady, the direct implication is that

$$\Sigma l_n P + \Sigma M_n u_n = (\text{constant}). \tag{7}$$

Evaluation of Equation (7) for initial conditions ahead of the wave and for conditions behind the wave produces

$$U \Sigma \overset{0}{M}_n = P \Sigma l_n + (U - \bar{u}) \Sigma M_n .$$

For convenience, the pressure ahead of the wave is assumed to be zero. Since for a unit of time, $\Sigma \overset{0}{M}_n = \Sigma M_n = U \overset{0}{L} \Sigma \overset{0}{\alpha}_n \overset{0}{\rho}_n$, it follows that

$$P = \bar{u}\, U \Sigma \overset{0}{\alpha}_n \overset{0}{\rho}_n . \tag{8}$$

Simultaneous solution of Equations (5) and (8) gives just

$$U = \sqrt{\frac{P}{(\Sigma \overset{0}{\alpha}_n \overset{0}{\rho}_n)\left(1 - \Sigma \dfrac{\overset{0}{\alpha}_n \overset{0}{\rho}_n}{\rho_n}\right)}} \tag{9}$$

This expression appears more familiar if one recognizes that the first quantity of the denominator is the average initial density of the composite and the second quantity is a measure of the strain. For a given pressure, the densities of the components are obtained from the hydrodynamic constitutive equations,

$$\rho_n = f(P) . \tag{10}$$

Thus, Equations (5), (8) and (10) permit the complete solution of the steady wave problem.

For steady wave propagation in the y-direction, symmetry conditions do not permit motion in the x-direction; consequently, uniaxial strain conditions are met exactly in each layer and it is sufficient to consider a unit area in the x-y plane. Initially,

$$\overset{0}{l}_n = \overset{0}{\alpha}_n L \tag{11}$$

and for all times

$$\Sigma\, l_n = L\,. \tag{12}$$

The repetition unit is no longer of constant length, $\overset{0}{L}$, but changes due to compression; hence, the time, t, for a *repetition length* to pass a point becomes

$$\begin{aligned} t &= \frac{\overset{0}{L}}{U} = \frac{\Sigma\, \overset{0}{l}_n}{U} \\ &= \frac{L}{U - \overline{u}} = \frac{\Sigma\, l_n}{U - \overline{u}} \end{aligned} \tag{13}$$

Balance of mass for the composite for flow into and away from the wave front per unit time becomes from Equation (13) just

$$U\,\Sigma\, l_n = (U - \overline{u}\,)\,\Sigma\, \overset{0}{l}_n\,. \tag{14}$$

Since $l_n = \dfrac{\overset{0}{l}_n \overset{0}{\rho}_n}{\rho_n}$ obtains for each component

$$U\,\overset{0}{L}\,\Sigma\,\frac{\overset{0}{\alpha}_n \overset{0}{\rho}_n}{\rho_n} = (U - \overline{u}\,)\,\overset{0}{L}$$

or

$$\overline{u} = U\left(1 - \Sigma\,\frac{\overset{0}{\alpha}_n \overset{0}{\rho}_n}{\rho_n}\right). \tag{15}$$

Balance of momentum follows directly as before. We take the mass per unit time as $U\,\Sigma\,\overset{0}{\alpha}_n\overset{0}{\rho}$ because $\overset{0}{L}$ may be taken as unity; hence, the momentum balance across the wave becomes

$$U^2\Sigma\,\overset{0}{\alpha}_n\overset{0}{\rho}_n = P + (U - \overline{u}\,)\,U\,\Sigma\,\overset{0}{\alpha}_n\overset{0}{\rho}_n\,.$$

This reduces to

$$P = \overline{u}\,U\,\Sigma\,\overset{0}{\alpha}_n\overset{0}{\rho}_n\,. \tag{16}$$

Simultaneous solution of Equations (15) and (16) results in

$$U = \sqrt{\frac{P}{(\Sigma\,\overset{0}{\alpha}_n\overset{0}{\rho}_n)\left(1 - \Sigma\,\dfrac{\overset{0}{\alpha}_n\overset{0}{\rho}_n}{\rho_n}\right)}} \tag{17}$$

where the densities of the components at a specified pressure are again obtained through Equation (10).

Comparison of Equations (9) and (17) clearly shows that steady wave analysis has led to identical results for the two extremes of laminate orientation. Although surprising at first, on closer examination the result is completely reasonable. Requiring hydrodynamic behavior for the individual components and neglecting dispersive effects suggest that structural details will disappear in the analysis. Indeed, this is substantiated by the geometric indifference of the theory. In application of the theory, we should remember that it approaches physical reality only at high pressure levels and far removed from the wave front. At the lower stress levels where real composites support shear stresses both within the layers and at layer interfaces, we expect the theory to give incorrect shock velocities. In general, the theory will give a lower bound in velocities.

The geometric indifference displayed by this theory suggests that it is applicable to a wider range of composites; i.e., mechanical mixtures. A similar conclusion, beginning from a mechanical mixture model, was arrived at by Tsou and Chou [6]. In particular, here we will compare the results of this theory with experimental data obtained on a mechanical mixture consisting of small particles imbedded in a soft matrix. However, since the theory requires only the volume fraction and the equation of state of the components, it must also provide prediction in fiber reinforced composites, where the assumption of geometric "fineness" is met.

DISCUSSION

Composites:

There have been several theories developed recently to predict the high pressure dynamic behavior of fibrous or laminated media and homogeneous mixtures. Although the individual models develop differently, some lead to basically the same results.

Two other variations using steady wave analysis have been advanced by Tsou and Chou [3] for fibrous composites and Torvik [7] for laminated composites. Both these analyses limit consideration to propagation parallel to the plates or fibers. Tsou and Chou's system of equations includes an energy equation and a specification of the final temperatures of the components. They have obtained solutions for what they refer to as "isothermal" and "adiabatic" conditions. Actually these conditions do not represent a thermodynamic loading path for a component but denote complete or no heat transfer between components.

Although the Tsou and Chou model was fibrous, the assumptions of one-dimensional steady wave propagation produce equations inherently free of

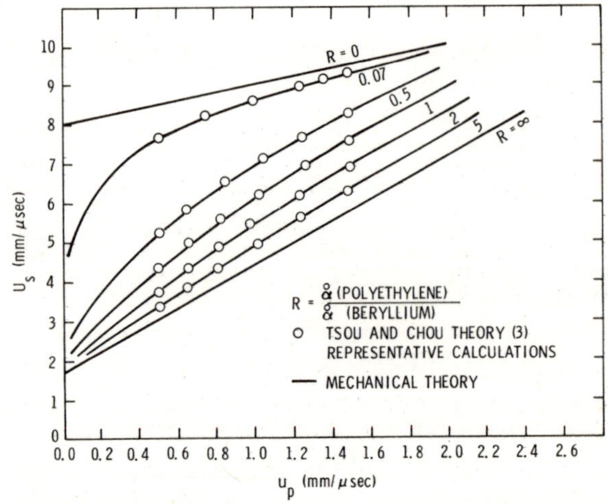

***Figure 2.** Comparison of theoretical calculations for hypothetical laminate.*

geometrically dependent terms, as we have illustrated previously. Consequently, we may compare their theory directly to the results of this study, as shown in Figure 2. We have based our calculations on the Hugoniots of the hypothetical beryllium-polyethylene components* as given by Tsou and Chou. For this composite, they found about a one percent difference between their two conditions, which makes the calculations indistinguishable on the plot. Our calculations, in turn, appear to give identical results to those of Tsou and Chou; even though the model is computationally much simpler. Perhaps with materials of widely different heat capacities, the "isothermal" calculations of Tsou and Chou would show significant deviations from the mechanical theory. However, for the fiber dimensions common in composites, there is some question whether sufficient time is available for significant heat flow between components.

Superficially, a theory which explicitly incorporates thermodynamics appears inherently more attractive than a mechanical theory; two facts mitigate this thought. First, correct thermodynamic treatment of the problem requires detailed specification of the loading path of the components. The actual loading path of the components is quite complex since each component achieves equilibrium through a series of wave reverberations and long term heat transfer. One can avoid these complexities, such as Tsou and Chou [3, 6] have done, by assuming that a certain thermodynamic end point is achieved.

---

* **Beryllium $U_s = 8.0 + 1.0\, u_p$ (mm/μsec); $\overset{0}{\rho} = 1.845$ g/cm³ and polyethylene $U_s = 1.68 + 2.75\, u_p$ (mm/μsec); $\overset{0}{\rho} = 0.91$ g/cm³, where U_s is the measured shock velocity and u_p the particle velocity, were used in the calculations.**

Second, implicit inclusion of certain elementary loading paths may be obtained more directly and far easier through the choice of the constitutive equations in the calculationally simple mechanical theory.

Torvik [7] has used a steady wave model based on a mechanical system of equations for propagation parallel to the laminates of a composite. A rather unusual assumption of complete decoupling between layers is made. Complete decoupling is a very strong restriction which is physically realistic only under very special circumstances. Torvik has also illustrated his model with a hypothetical composite; in this case, aluminum-polymethyl methacrylate. No comparison of our calculations may be made with his theory because a unique particle velocity is not obtained for his model. However, Torvik presents two simplified functions for calculating an "average" particle velocity. Using precisely the same Hugoniot relationships, we may compare our results to Torvik's simplified theory. The comparison in Table 1 shows marked differences in predicted shock and particle velocities. The differences are especially noticeable in the particle velocities.

A recently reported effective modulus analysis by the author [8] leads to precisely the same results as the steady wave analysis. Therefore, in reality these two theoretical approaches are strongly inter-related.

Mixtures:

Numerous simple mixture theories to predict the Hugoniot response at high pressure levels have been proposed. The early theories, as summarized

Table 1. Shock Parameters for Hypothetical Aluminum-Polymethyl Methacrylate (PMMA) Composite, Comparison to Torvik Model.

			TORVIK MODEL						
			NORMAL			SIMPLIFIED‡		THIS WORK	
α^0_{Al}	α^0_{PMMA}	P kbar	U^*_s mm/μsec	u_{pAl} mm/μsec	u_{pPMMA} mm/μsec	u_{pe} mm/μsec	$u_{pe'}$ mm/μsec	U_s mm/μsec	u_p mm/μsec
.9	.1	15.7	4.65	.126	.278	.133	.140	4.75	.126
		75.7	5.59	.505	1.084	.531	.557	5.53	.523
		162.7	6.46	.935	2.071	.988	1.043	6.30	.987
.5	.5	15.7	3.66	.161	.357	.221	.257	3.61	.191
		75.7	5.03	.568	1.246	.775	.894	4.66	.713
		162.7	6.30	.965	2.162	1.330	1.554	5.65	1.263
.1	.9	15.7	3.37	.176	.393	.349	.371	3.25	.262
		75.7	4.84	.593	1.315	1.169	1.239	4.37	.939
		162.7	6.24	.976	2.198	1.951	2.073	5.47	1.613

* U_s is the same for both the normal and simplified model.
‡ u_{p_e} and $u_{p_{e'}}$ are from Eqs. (21) and (22) of Reference 7.

Hugoniot of Components
Aluminum $U_s = 5.27 + 1.438\ u_p$; $\rho_0 = 2.70$ gm/cm^3
PMMA $U_s = 2.7 + 1.61\ u_p$; $\rho = 1.86$ gm/cm^3

by Chabai [9], hypothesize a mixing rule from which various properties of the mixture are calculated. For example Chabai selected a variation of Dalton's law to calculate the molar partial pressures of the elemental constituents of the mixture at constant volume. In most instances, the form of the mixing rule differs substantially from that derived by steady wave analysis and may be physically unrealistic for mechanical mixtures of solid components. The steady wave analysis yields an equivalent of a mixing rule which, at a constant pressure, weight component densities by initial volume fractions. This mixing rule is identical to that assumed by McQueen *et al* in their recent work [10]. Because of differences in accounting for energy, however, the steady wave analysis will give identical results to McQueen *et al* calculations only for the zero degree Kelvin isotherm. In many respects, the Tsou and Chou [6] mixture derivations appear similar to those of McQueen *et al.* and to the results of this paper.

The steady wave results for mixtures are identical to an analysis by Torvik [11]* and to an effective modulus analysis given recently by the authors [8], as previously stated, and to the earlier effective modulus analysis of Dremin and Karputkhin [13]. Dremin and Karputkhin utilized the analysis in a novel manner to deduce the unknown Hugoniot of one component of a geologic material where the mixture Hugoniot and all other component Hugoniots were known.

Use of the low pressure limit of the steady wave analysis to predict acoustic velocities has been proposed by the authors [8] and Torvik [11]. However, the limitation to hydrodynamic behavior may be so severe as to negate the usefulness of this approach. In most practical cases, the fact that the material supports shear may predominate at low pressures. Along this line, Budiansky [14] has developed an analysis for a spherical inclusion in an isotropic matrix using linear elasticity theory. This analysis reduces to Equation (17) in the limit as the shear moduli of the components vanish; i.e., as they behave hydrodynamically.

Comparison to Experiment:

We shall draw our comparisons first from stress wave profile results of an unusual experiment using stainless steel-epoxy plate laminates and then from results of Hugoniot experiments. The plate laminate data are from one transmitted wave experiment performed at a stress level of a few kilobars. While the stress level is not really compatible with the hydrodynamic constraints of the model, the comparison should clearly define the influence of the approximations used in developing the model. Except for two systems,

* The derivations of steady wave analysis for mixtures appeared in internal reports nearly simultaneously. [11, 12]

Hugoniot data on non-porous composites and mechanical mixtures are meager. Although these two systems, a filament wound quartz-phenolic composite and a particulate Al_2O_3-epoxy mixture, are not completely explored, sufficient information exists to give a meaningful comparison. Both examples are non-ideal fluids; consequently, we expect to observe deviations between theory and experiment. Selection of component constitutive relationships which conform closely to the hydrodynamic restriction is a significant problem.

Stainless Steel-Epoxy:

Recently Lundergan and Drumheller [15] reported the results of plate impact experiments on a 304 stainless steel-Epon 828 epoxy plate laminate. These specimens were impacted by low velocity projectiles and the transmitted particle velocity profile was measured using a velocity interferometer technique. Wave propagation was perpendicular to the plane of the laminates. In most of these experiments, the impactor produced an attenuating short duration pulse and cannot be analyzed by the current theory since it specifies only the compression path. However, one experiment which used a thick impactor to produce an experimentally infinite compressive pulse can be analyzed. The test configuration was a 11.08 mm thick Epon 828 impactor with an initial velocity of 0.1748 mm/μsec and a 10.09 mm thick composite specimen composed of 10 bilaminate plates. Steel comprises 0.7552 of the composite volume. The specimen is followed by a 1 mm thick buffer of Epon 828 and a very thick polymethyl methacrylate interferometer window. Impedances of the epoxy and the polymethyl methacrylate are essentially identical. The particle velocity was measured at the buffer/window interface.

For the austenitic 304 stainless steel, we take the dynamic hydrostat of $U_s = 4.316 + 1.89\, u_p$ (mm/μsec); $\overset{0}{\rho} = 7.896$ g/cm^3 from Fowler *et al* [16]. For the Epon 828, as well as the window material, we use the equilibrium response of Epon 828 with hardener Z as determined by Guess [17]. Polymeric materials are viscoelastic, displaying both instantaneous and equilibrium responses in one-dimensional loadings [18]. In our analysis, it was necessary to use the one-dimensional strain equilibrium curve. The difference between this curve and the equilibrium hydrostat is on the order of the deviatoric component of the stress field, which is small. The equilibrium uniaxial strain curve is given by $U_s = 2.64 + 1.66\, u_p$ in mm/μsec; $\rho_o = 1.2$ g/cm^3.

The results of the calculation are compared to the experimental wave profile in Figure 3. This comparison shows distinctly the influence of the Hugoniot assumptions; i.e., the theory gives a single sharp shock front and does not exhibit the smearing of the front nor the oscillations which result from the dispersive nature of the composite. It is seen that the theory gives a good first approximation to the experimental observation. The calculated

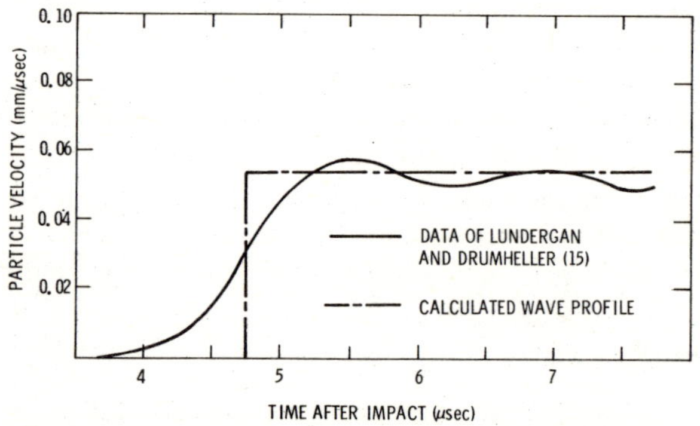

Figure 3. *Predicted and experimental wave profiles in a laminate plate composite of stainless steel and epoxy plates.*

arrival time corresponds very well to the measured half amplitude arrival time and the calculated particle velocity is very close to the average of the experimental oscillations in particle velocity, as expected.

Quartz-Phenolic:

Experimental Hugoniot data on filament wound quartz-phenolic are available from the work of Lee [19] and May and Kinsey [20]. The 70 weight percent quartz material had an initial density of 1.85 g/cm³ and was processed to minimize porosity. Density of the filament material is consistent with fused quartz, but higher, 2.29 g/cm³, perhaps because of experimental errors in density measurements on filaments. The Hugoniot data are consistent with other determinations on similar quartz-phenolic materials [21, 22]. We have not used these data in the comparison because differences in experimental techniques and sample fabrication cause an apparent scatter of data. This apparent scatter is avoided by selecting data from a single group of experimenters.

Hugoniot data for the components were compiled from the isothermal compression data of Bridgman [23, 24] for fused quartz and the equilibrium response for epoxy, Epon 828 with hardener Z, of Guess [17]. A dynamic hydrostat [4] for fused quartz was calculated* from the isothermal data which included the energy contribution of the deviatoric and spherical portions of the stress field but does not include the deviatoric component of the stress. The calculations may lead to small energy errors and the curve fitting methods lead to small slope errors at the origin. These errors are probably

* We have assumed $\gamma\rho$ = constant and taken the Gruneisen coefficient, γ, as 0.9, the heat capacity, C_v, as .220 cal/g °C. From Wackerle's data[25] an average yield strength, σ_y, is taken as 30 kbar.

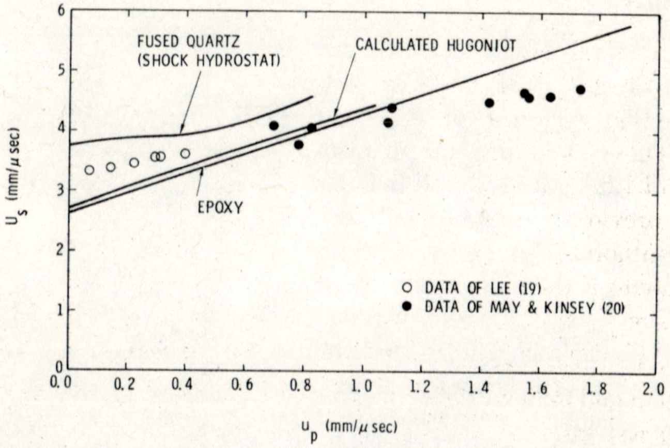

***Figure 4.** Quartz-phenolic Hugoniot data and hydrodynamic prediction.*

insignificant. From the calculations, we obtain a dynamic hydrostat of $P = 326\,\epsilon + 723\,\epsilon^2 - 10759\,\epsilon^3 + 64312\,\epsilon^4$; $0 \leqq P < 98$ kbar where ϵ is strain. At 98 kbars, fused quartz undergoes an abrupt change in behavior which may be either an elastic failure or a phase change [25]. However, we do not try to extend the analysis above 98 kbar in this paper.

A reliable Hugoniot for bulk phenolic is not available at this time; consequently, the equilibrium Hugoniot is taken for a comparable polymer, epoxy.

Calculated hydrodynamic behavior of the composite using the steady wave analysis is compared to the experimental data in Figures 4 and 5.* As Figure 4 indicates, the theoretical curve deviates considerably at low particle velocities from the experimental results. However, the stress magnitude deviation as shown in Figure 5 is quite small. Undoubtedly, this deviation is a measure of material strength effects. However, the close agreement between the calculated behavior and experiment suggests that the mechanical theory is adequate for approximate or engineering computations even to low stress levels.

The analysis is not carried above 98 kbars because of the change in the quartz response. If this is caused by a phase change, then the dynamic hydrostat of fused quartz would have to reflect the phase change as well as any thermal effects on the location of the phase line. If it is caused by elastic failure, then no appreciable modifications of the dynamic hydrostat result. Until the precise nature of fused quartz is known, the analysis cannot be extended above this point.

---

* We use a $U_s - u_p$ plot because it is particularly sensitive and a $P - \epsilon$ plot because it illustrates stress deviations clearly; ϵ is strain.

Al_2O_3-Epoxy:

Experimental Hugoniot data are available on 0.43, 0.34, and 0.20 volume fraction mixtures of Al_2O_3 in epoxy. These data are taken from Kent *et al* [26], Munson and Schuler [8] and previously unpublished work of the authors. The epoxy used in these systems is the same as that studied by Guess [17]. A pressure range of 0-150 kbars was studied for the 0.43 system and 0-20 kbars for the 0.34 and 0.20 systems.

A dynamic hydrostat for Al_2O_3 was obtained from the data of Munson [27], Ahrens *et al* [28] and McQueen and Marsh [29]. This is given by the equation $U_s = 8.14 + 1.28\ u_p$; $\rho_o = 3.969$ g/cm^3. A one-dimensional equilibrium curve, as previously described, was taken for the epoxy.

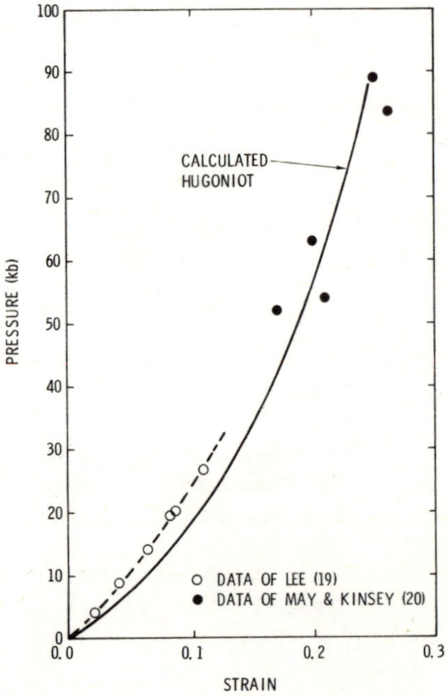

***Figure 5.** Quartz-phenolic pressure-strain data and hydrodynamic prediction.*

Predicted and experimental results are shown in Figures 6 and 7. In Figure 6, the experimental data show a clear deviation from the hydrodynamic prediction. The deviation is especially pronounced at low particle velocities in the 0.43 volume fraction Al_2O_3 data shown. This behavior is suggestive of the deviation caused by material strength. Although, the details of the deviation do not appear as normal elastic-plastic yielding behavior. The curves in Figure 7 show two distinct trends. The stress deviation between theory and experiment appears to increase smoothly with increasing strain* and appears to decrease with decreasing volume fraction of Al_2O_3. These trends are physically reasonable. The highest volume fraction material contains the greatest oxide volume which can be added, practically, to the epoxy and all particles are thought to be in contact. Although the soft epoxy matrix takes most of the volumetric strain, it offers very little resistance to shear. The apparent shear strength is predominantly the result of resistance to brittle failure of the Al_2O_3 particle contact points; directly analogous to the crushing of a porous solid [30]. Because the relative volume fractions of Al_2O_3 and epoxy change progressively

* This observation is true for 0.43 volume fraction Al_2O_3 only for strains less than approximately 0.15.

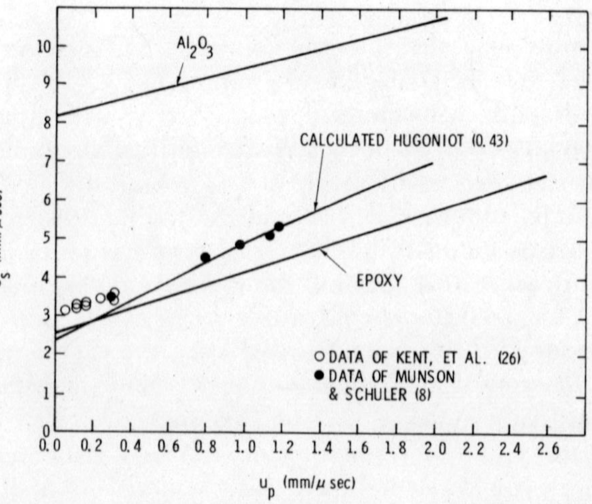

Figure 6. *Predicted and experimental Hugoniots, 0.43 volume fraction Al_2O_3-epoxy.*

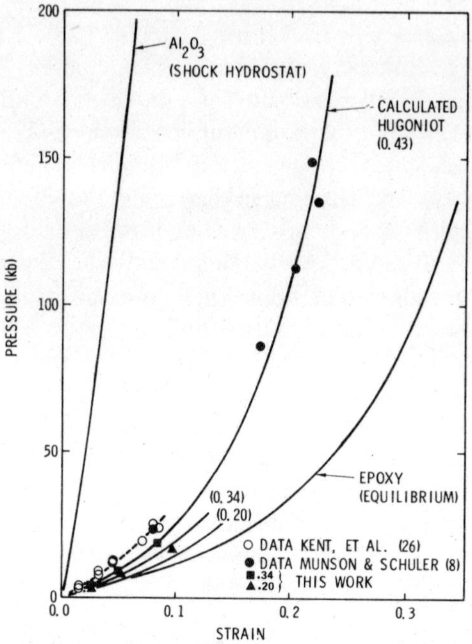

Figure 7. *Pressure-strain relationships and data for several Al_2O_3-epoxy mixtures.*

with pressure such that the fraction of Al_2O_3 increases, this implies that the contact area of Al_2O_3 increases progressively with strain; hence, the material strength will increase progressively with strain. Presumably, if the individual Al_2O_3 particles were completely decoupled, then an abrupt change in strength would occur at 0.43 volume fraction, with no apparent strength for mixtures below 0.43 volume fraction. However, the decoupling probably occurs gradually so that the stress deviation will decrease gradually also. In support of this concept, the theory and experiment come into close agreement for the 0.20 material, as would be expected.

At pressures of approximately 150 kbar, the experimental data appear to cross the theoretical prediction. Although scatter of data may produce such artifacts, we believe that it may be a consequence of the assumptions on the loading path; and, if the range of experimental data were larger, the trend would be clearly verified. Clearly, the components of the mixture will not load along the Rayleigh line, although this is a requirement of the Hugoniot calculations. We may make the following *thought experiment* to give an approximation to the path for an individual component of the mixture. We imagine a piece of component material, either matrix or reinforcement, impacted between two pieces of material which has the same equation of state as the prediction for the mixture. In this configuration, the component material will "ring into" the final pressure along a multiple shock path.

The results of this calculation are given in Table 2 for a final pressure of 205 kbar. Perhaps unexpectedly, the soft component loading path tends toward the adiabat. Hence, the soft component will undergo more strain and contain less energy than the comparable Hugoniot. The hard component, on the other hand, shocks to progressive points of higher pressures which causes the strain to be less and the internal energy to be greater than the Hugoniot. These contributions at high pressures *may* become significant and, since the soft component influences the behavior strongly, the mixture curve will in general fall below the calculations which assume a single shock loading path for the components.

Table 2. Multiple Shock Calculations for a Final Pressure of 205 kb.

	Single Shock		Multiple Shock	
Material	**Strain**	**Energy (cal/g)**	**Strain**	**Energy (cal/g)**
Epoxy	.369	3.18	.378	2.39
Al_2O_3	.0654	.169	.0639	.184
Mixture*	.238	1.03	.243	.815

* Calculated from volume fraction weightings of strain and mass fraction weightings of the component energies.

Although it is difficult to know the precise loading path, the above approximation gives some insight into the problem. In fact, the insight may produce an average loading path which is quite good for random mixtures. For regular geometric arrays, however, one may not be able to use the concept.

CONCLUSIONS

A steady wave analysis has been developed for the prediction of constitutive equations of composites and mixtures at high pressures from the constitutive equations of the components. This simple hydrodynamic analysis yields identical results for wave propagation parallel and normal to the layers of an *n*-layer composite. Consequently, the theory was extended directly to mechanical mixtures.

Comparison of the calculations show an excellent first approximation to the experimentally determined wave profile of a plate laminate composite. In particular, the model predicts accurately the half amplitude propagation velocity and the final particle velocity of the wave.

Comparison of the results of experiment to the calculated Hugoniots of a filament wound quartz phenolic composite and three aluminum oxide-epoxy mixtures is quite good. Minor deviations occur between experiment and theory at low stresses because of mechanical strength effects which are not predictable from hydrodynamic analyses. Since the strength or deviatoric component of the stress is generally small compared to the hydrodynamic or spherical component, the theory gives a close approximation to real behavior. At large stresses, we expect on theoretical grounds that the energy variations between the simple loading path assumed in our calculations and the actual loading path will cause deviations, which will grow progressively larger as the stress level increases. However, thermodynamic deviations appear to be comparatively insignificant within the range of experimental data used for comparison in this report.

The theory is related to other comparable theories using steady wave and effective modulus analyses. In general, many of these theories are directly related to each other although they were originally derived from various physical models of wave propagation in composites. An important aspect of all these theories whether they originate from models of mixtures or fibrous and plate composites is that they yield similar results. This is a consequence of the overriding influence of the assumption of hydrodynamic component behavior compared to the minor influence of the details of the geometric model or of the energy differences between the various *simple* loading paths implied by the models. In some instances, the theory developed here yields Hugoniots in a calculationally much simpler manner than other models.

ACKNOWLEDGMENTS

The authors wish to thank L. Lee for use of unpublished data. The Al_2O_3-epoxy specimens were especially prepared by A. Quant, for which we are very appreciative. We are indebted to J. Enri and C. Witten for experimental support and to M. Percival and H. Sutherland for discussions of the manuscript.

REFERENCES

1. C. T. Sun, J. D. Achenbach, and G. Herrmann, "Continuum Theory for a Laminated Medium," *J. Appl. Mech.,* Vol. 35 (1968), p. 467.
2. J. C. Peck and G. A. Gurtman, "Dispersive Pulse Propagation Parallel to the Interfaces of a Laminated Composite," *J. Appl. Mech.,* Vol. 36 (1969), p. 479.
3. F. K. Tsou and P. C. Chou, "Analytical Study of Hugoniot in Unidirectional Fiber Reinforced Composites," *J. of Composite Materials,* Vol. 3 (1969), p. 500.
4. D. E. Munson and L. M. Barker, "Dynamically Determined Pressure-Volume Relationships for Aluminum, Copper, and Lead," *J. Appl. Phys.,* Vol. 37 (1966), p. 1652.
5. W. Herrmann, "Theory of Motion in One Dimension I, Balance Relations for a Single Material," SC-TM-68-648, Sandia Laboratories, Albuquerque, N. Mex. (1968).
6. F. K. Tsou and P. C. Chou, "The Control-Volume Approach to Hugoniot of Macroscopically Homogeneous Composites," *J. of Composite Materials,* Vol. 4 (1970), p. 526.
7. P. J. Torvik, "Shock Propagation in a Composite Material," *J. of Composite Materials,* Vol. 4 (1970), p. 296.
8. D. E. Munson and K. W. Schuler, "Hugoniot Predictions for Mechanical Mixtures using Effective Moduli," Sagamore Army Materials Conference, held Sagamore, N. Y., September 1970, to be published.
9. A. J. Chabai, "Synthesis of Shock Hugoniots for Rock Materials," *Rock Mechanics,* Proceedings of 5th Symposium on Rock Mechanics, University of Minnesota, ed. C. Fairhurst, Pergamon Press, Inc. (1963).
10. R. G. McQueen, S. P. Marsh, W. J. Carter, J. N. Fritz and J. W. Taylor, "The Equation of State of Solids," *Shock Wave Studies in Hypervelocity Impact Phenomena,* ed. E. Cohen, Academic Press, Inc. (1969).
11. P. J. Torvik, "A Simple Theory for Shock Propagation in Homogeneous Mixtures," AFIT TR 70-3, Air University, Wright-Patterson Air Force Base, Ohio (1970).
12. D. E. Munson and K. W. Schuler, "Hugoniot Predictions in Laminated Composites and Mechanical Mixtures," SC-DR-70-34, Sandia Laboratories, Albuquerque, N. Mex. (1970).
13. A. N. Dremin and I. A. Karputkhin, "Method of Determining the Shock Adiabats of Dispersed Substances," *Zh. Prik. Mekh. i. Texh, Fiz.,* Vol. 3 (1960), p. 184.
14. B. Budiansky, "Thermal and Thermoelastic Properties of Isotropic Composites," *J. of Composite Materials,* Vol. 4 (1970), p. 286.
15. C. D. Lundergan and D. S. Drumheller, "Propagation of Stress Waves in a Laminated Plate Composite," *J. Appl. Physics,* Vol. 42 (1971) in press.
16. C. M. Fowler, F. S. Minshall, and E. G. Zukas, "A Metallurgical Method for Simplifying the Determination of Hugoniot Curves for Iron Alloys in the Two-Wave Region," *Response of Metals to High Velocity Deformation,* Interscience Publishers (1961).
17. T. R. Guess, "Some Dynamic Mechanical Properties of an Epoxy," SC-RR-68-343, Sandia Laboratories, Albuquerque, N. Mex. (1968).
18. J. W. Nunziato, K. W. Schuler and E. K. Walsh, "On the Bulk Response of Nonlinear Viscoelastic Solids," SC-DR-70-630, Sandia Laboratories, Albuquerque, N. Mex. (1970).
19. L. M. Lee, Sandia Laboratories, Albuquerque, N. Mex., unpublished data.

20. R. P. May and C. H. Kinsey, "Preliminary High Pressure Data for Filament Wound Quartz Phenolic (FWQP)," SC-TM-67-642, Sandia Laboratories, Albuquerque, N. Mex. (1967).
21. W. M. Isbell, N. H. Froula, F. H. Shipman, "Shock Wave Propagation and Equation of State Measurements of Quartz Phenolic," *Fundamental Material Behavior Study,* Vol. III, BSD-TR-67-25, Ballistic Systems Division, Norton AFB, California (1967).
22. J. A. Charest and B. D. Jenrette, "Hugoniot Equations-of-State of Tape Wrapped Two-Dimensional Quartz Phenolic," Tech. Rpt. No. S-448-R, EG&G, Santa Barbara, California (1969).
23. P. W. Bridgman, "The Compression of 39 Substances to 100,000 kg/cm^2," *Proc. Am. Acad. Arts Sci.,* Vol. 76 (1948), p. 55.
24. P. W. Bridgman, "Rough Compressions of 177 Substances to 40,000 kg/cm^2," *Proc. Am. Acad. Arts Sci.,* Vol. 76 (1948), p. 71.
25. J. Wackerle, "Shock-Wave Compression of Quartz," *J. Appl. Phys.,* Vol. 33 (1962), p. 922.
26. L. Kent, K. J. Bowen, and B. M. Butcher, "The Performance of Manganin Wire Gages in Aluminum Oxide Filled Epoxy," SC-DR-70-291, Sandia Laboratories, Albuquerque, N. Mex. (1970).
27. D. E. Munson, "Dynamic Mechanical Behavior of Polycrystalline Al_2O_3," SC-RR-69-803, Sandia Laboratories, Albuquerque, N. Mex. (1970).
28. T. J. Ahrens, W. H. Gust and E. B. Royce, "Material Strength Effect in the Shock Compression of Alumina," *J. Appl. Phys.,* Vol. 39 (1968), p. 4610.
29. R. G. McQueen and S. P. Marsh, "Compressibility: Elastic Constants (F. Birch)," *Handbook of Physical Constants,* ed. S. P. Clark, Jr., Geological Society of America, New York (1966).
30. W. Herrmann, "Constitutive Equation for Dynamic Compaction of Ductile Porous Materials," *J. Appl. Phys.,* Vol. 40 (1969), p. 2490.

Fatigue Crack Propagation in a Discontinuous Composite

PETER A. THORNTON, *Benet R&E Laboratories, Watervliet Arsenal*
Watervliet, N. Y. 12189

(Received December 1, 1971)

Fatigue crack propagation in fiber reinforced composites is rapidly becoming an important aspect of materials-science research. This is due, chiefly, to the increasing structural application of these materials, accompanied by conditions of cyclical loading which may subject the composite to subcritical crack growth. Since this area has been well documented for metals and several recent investigations [1,2] have indicated that increased crack growth resistance can be obtained in polymeric materials, an attempt was made to evaluate the fatigue crack propagation characteristics of an epoxy resin matrix reinforced with discontinuous metal fibers.

Single edge notched specimens (SEN), 2″ wide, 8″ long and ½″ thick were cast from the below listed system and cured at room temperature. Fibers were randomly oriented with an average volume ratio of 0.134 (13.4% ± 1.0). The average ultimate tensile strengths of the matrix and composite were 7790 psi and 4742 psi, respectively.

Matrix	**Fiber**
Shell Epon 828	Aluminum
Versamid 140	.005″ x .005″ x .125″
(60 resin/40 hardener)	

These bars were pin loaded, with an interference fit at the grips and tested in zero-tension-zero loading on a Sonntag Fatigue Testing Machine at a frequency of 30 cycles/sec. Loads of 600# and 800# were applied and crack growth monitored by a traveling microscope.

Macroscopic crack growth rate, *da/dn,* was related to the stress intensity at the crack tip by the expression [3]:

$$da/dn = c(\Delta k)^m$$

where c = constant; Δk = stress intensity factor range; m = exponent (approximately = 2-6 for metals). Stress intensity for this specimen geometry was obtained from the following relationship [4]:

$$k = \sigma a^{1/2} \, \{5\pi^{1/2} \, [20-13(a/w)-7(a/w)^2]^{-1/2}\}$$

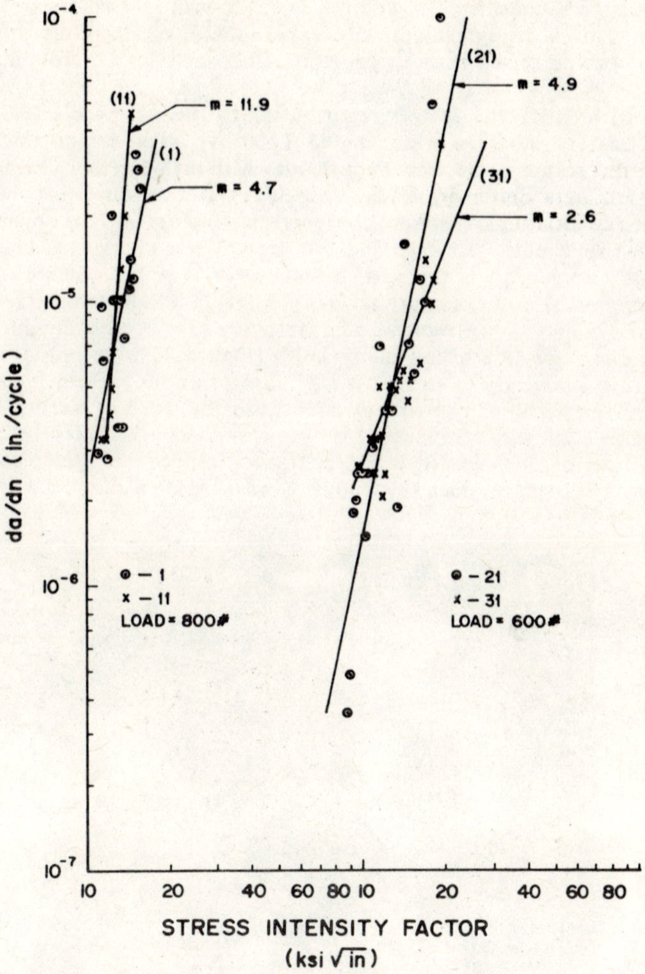

Figure 1. Log growth (da/dn) vs. log stress intensity factor range (ΔK).

This expression is valid for specimens without rotation and ($a/w<0.6$), where a = total crack length and w = width of the bar. This analysis was also subject to the condition that the distance between loading points > $3w$.

Fatigue crack propagation rates from 3.6×10^{-7} in./cycle to 1.0×10^{-4} in./cycle, and stable crack lengths up to 1.2" were measured for this material. These data (summarized below), along with the curves from a linear regression analysis are presented in Figure 1.

Specimen No.	*Load (#)*	*m*	*r(correlation coeff.)*
1	800	4.7	0.67
11	800	11.9	0.98
31	600	2.6	0.72
21	600	4.9	0.88

It is apparent from the slopes "m" (change in crack growth rate) of the straight line approximations that considerable variation in crack propagation behavior exists for this material. This is in agreement with experimental observations where the advancing crack was perceived to both penetrate fibers and deviate around them (delamination).

In order to identify the factors contributing to this observed crack growth behavior, the fracture surfaces were coated (200 Å) with carbon and gold, and examined in the scanning electron microscope. Although fracture characteristics in the matrix itself were similar for all bars, viz., flat, rather featureless cohesive failure and some fatigue striations, distinct differences in fracture surface appearance were found between specimens with high and low growth rate change (as depicted by m) for the respective loads. In the case of relatively large m, shallow cavities (air bubbles) were present in the fracture surface (Figure 2). Qualitatively, the percentage of fibers in the main crack plane was comparatively low – probably due to segregation or "bundling" of fibers, and many holes (Figure 3) were noted where fibers pulled out. This is contrary to current energy absorption theory which contends that the process of fiber pullout contributes directly to the work of fracture and, hence, to the toughness of the composite [5]. However, the geometry of these holes suggests poor wetting between fiber and matrix (entrapped air), which could explain the lower crack resistance. Just the opposite was observed in specimens with rela-

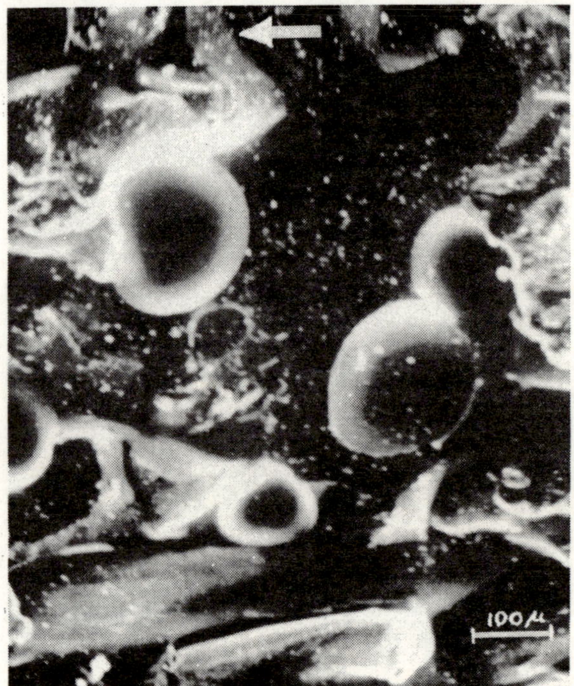

Figure 2. Fracture surface containing air bubble cavities, arrow denotes main crack direction.

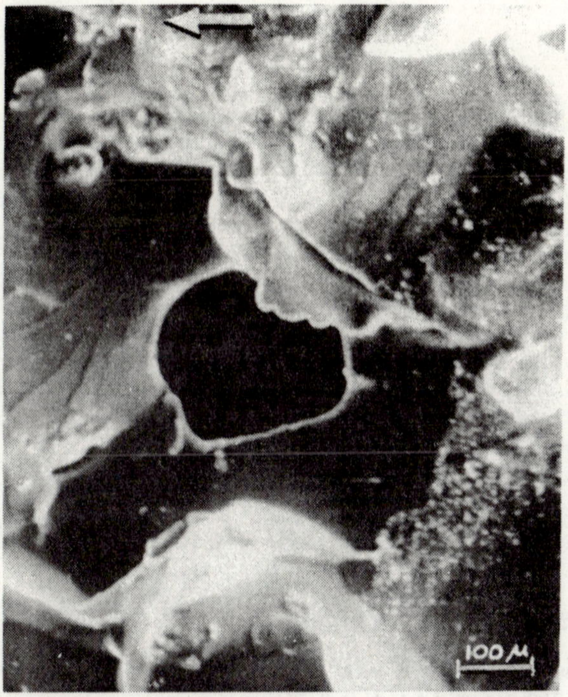

***Figure 3.** Deep hole in fracture surface resulting from fiber pull-out, arrow denotes main crack direction.*

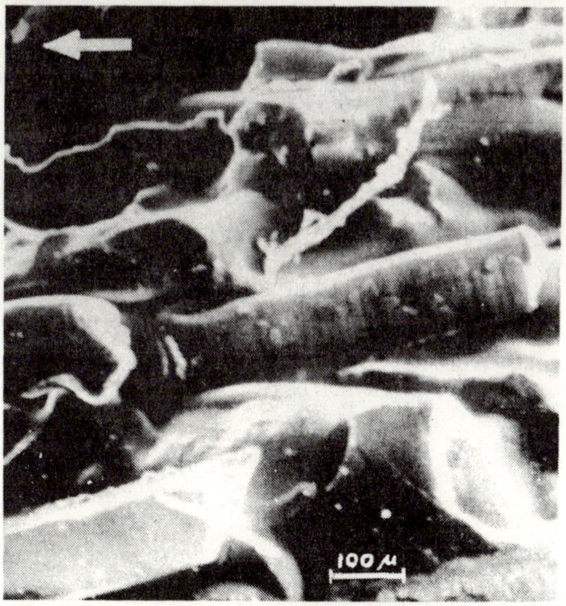

***Figure 4.** Fractograph showing fibers in main crack plane and good wetting characteristics, arrow denotes main crack direction.*

tively small *m* values. Fracture surfaces contained very few holes, indicating good wetting, and practically no air bubble cavities. A much greater percentage of fibers was associated with the fracture surface and they generally appeared to be oriented at small angles to the main crack plane (Figure 4). This photograph also displays the good wetting characteristics that were present in the low *m* specimens (traces of resin remain in contact with fiber).

SEN specimens of the unreinforced matrix were similarly tested. Yet at loads as low as 100#, all bars failed catastrophically with virtually no stable crack growth. Therefore, it is apparent that significant toughening can be obtained with discontinuous fibers even at low volume fractions (13%). Also, a correlation can be determined between crack propagation rate and stress intensity at the advancing crack tip, (ΔK) in this material. Furthermore, this correlation and the change in crack growth rate as a function of ΔK appears to be very sensitive to fabrication and processing parameters.

ACKNOWLEDGMENT

The assistance of the Watervliet Arsenal Organic Composite Laboratory personnel, and Mr. W. Mortimer, are sincerely appreciated.

REFERENCES

1. R. W. Hertzberg, H. Nordberg and J. A. Manson, "Fatigue Crack Propagation in Polymeric Materials", *J. Mat. Sci.*, Vol. 6 (1970), p. 521.
2. F. J. McGarry, "Crack Propagation in Fiber Reinforced Plastic Composites", MIT Research Report P66-5, October 1966.
3. P. C. Paris, "The Fracture Mechanics Approach to Fatigue" *Fatigue An Interdisciplinary Approach,* Proc. 10th Sagamore Army Material Research Conference, Syracuse University Press, (1964).
4. L. P. Pook, "The Effect of Friction on Pin Jointed Single Edge Notch Fracture Toughness Test Specimens", *Inter. J. Fracture Mech.*, Vol. 4 (1968), p. 295.
5. G. A. Cooper, "The Fracture Toughness of Composites Reinforced with Weakened Fibers", *J. Mat. Sci.*, Vol. 5 (1970), p. 645.

Structure-Property Relations and Reliability Concepts

JOHN C. HALPIN

Air Force Materials Laboratory
Dayton, Ohio 45433

(Received February 28, 1972)

ABSTRACT

Typical procedures currently employed in engineering design are reviewed and applied to the evaluation of composite technology as it exists at this time. Special attention is directed to the topics of stiffness, static strength, specific weight, fatigue, structural reliability and adhesive joining. Verification of the achievement of a safe composite design by a proof-test concept is introduced. Evaluation of composites vis-a-vis metallic technology yields a positive attitude toward the future of composite technology. Significant support is found to justify the long-term potential of advanced filamentary composites.

INTRODUCTION

THE DECADE OF the Fifties opened with an uninhibited enthusiasm for the extraordinarily high strength and stiffness properties primarily available in single-crystal whiskers but also in continuous filament forms. While the realization of the potential of defect-free "whiskers" in engineering composites has seeped away in the murk of technological reality, striking progress has been achieved in advanced filament reinforcement. The Sixties witnessed the development of boron and graphitized polymeric fibers as well as several other potential reinforcements. In fact, the technology has developed to the point that "proof-of-concept" is now being demonstrated in operational engineering structural items [1, 2]. At this point, it is interesting to note that our understanding of composite solids has altered in a subtle but nonetheless radical manner. The initial emphasis upon netting analysis, high fiber properties, has been tempered by the realization that the engineer cannot, nor will not, execute design based solely upon longitudinal filament properties. Technical composites are multiphased, anisotropic bodies. Efficient and reliable use of such materials necessitates optimization of a laminated composite with respect to all of its directional properties (particularly, transverse tensile shear, and longitudinal compression, as well as the earlier emphasis upon longitudinal properties).

Employing current design engineering characterization and design procedures, we are in a position today to evaluate the performance capability of composites for a variety of filaments ranging from E-glass through the very high-modulus, high-strength graphite systems.

LAMINATE PROPERTIES – STRENGTH AND STIFFNESS

Current attitudes regarding composite materials [3, 4] emphasize the relationship of structural performance to the properties of a ply. A "ply" is a thin sheet of material consisting of an oriented array of fibers embedded in a continuous matrix material. These plies are stacked one upon another, in a defined sequence and orientation, and bonded together yielding a laminate with tailored properties. The properties of the laminate are related to the properties of the ply by the specification of the ply thickness, stacking sequence, and the orientation of each ply. The properties of the ply are, in turn, specified by the properties of the fiber and matrix, their volumetric concentration, and geometric packing in the ply. Generally, the ply material is preformed and can be purchased in a continuous compliant tape or sheet form which is in a chemically semi-cured condition. Fabrication of structural items involves using this "prepreg" material and either winding it onto a mandrel or cutting and stacking it onto a mold after which either heat and pressure or tension is applied to complete the chemical hardening process.

The basis for engineering design for such a material is then – the properties of a cured ply or lamina *as it exists in a laminate.* For example, consider the ply illustrated in Figure 1. This ply is treated as a thin, two-dimensional item and is mechanically characterized by its stress-strain response to loadings in: (1) the direction of the filaments which exhibit a nearly linear response up to a rather large fracture stress; (2) in the direction transverse to the filament orientation which exhibits a significantly decreased moduli and strength; and (3) the response of the material to an in-plane shear load not illustrated in Figure 1.

By contrast with isotropic metallic materials, an oriented ply, in the form of a thin sheet, is anisotropic and requires four elastic (plane stress) constants [3, 4] to specify its stiffness properties in its natural orientation:

$$\begin{aligned}\sigma_1 &= Q_{11}\epsilon_1 + Q_{12}\epsilon_2\\ \sigma_2 &= Q_{12}\epsilon_1 + Q_{22}\epsilon_2\\ \sigma_6 &= Q_{66}\epsilon_6\end{aligned} \qquad (1)$$

The elastic stress-strain properties of a laminate are related to the ply properties through the assumptions:

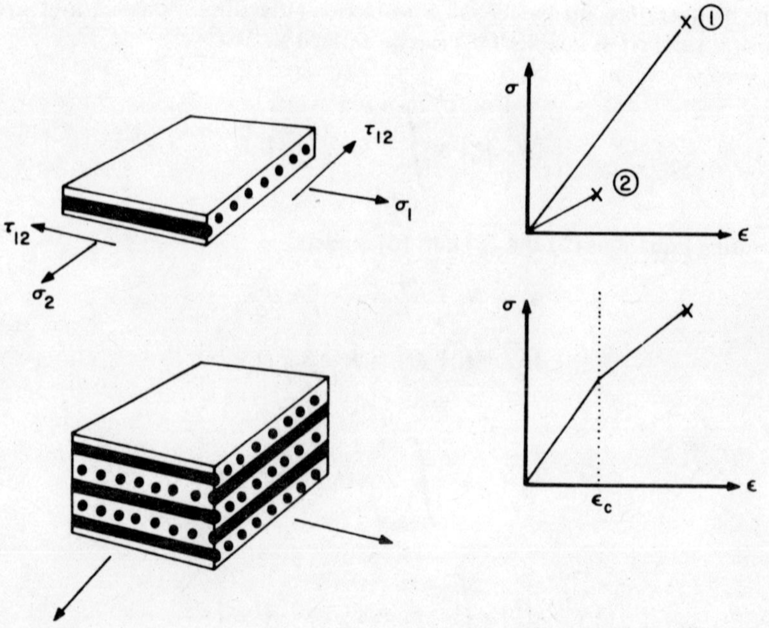

Figure 1. Illustration of lamina characterization and strength properties at a laminate.

1) All layers are in a state of plane stress; the stress transverse to lamination plane are neglected;

$$\sigma_3 = \sigma_4 = \sigma_5 = 0 \tag{2}$$

2) All layers are bonded together and the strain components, in the 1-2 plane are linear functions of z, the distance through the depth of the laminate;

$$\epsilon_i = \epsilon_i + z\, k_i \tag{3}$$

where ϵ°= in-plane or mid-plane stretching and k=plate curvature, $i = 1, 2$ refers to the normal components, and $i = 6$ refers to the shear component, the engineering strain component.

3) All layers obey generalized Hooke's law; (see Equation 1). The stress-strain relation for the assumed plane-stress condition, including the thermal effect for each layer is:

$$\sigma_i = Q_{ij}\,(\epsilon_j - \alpha_j\, T), \quad i,j = 1,2,6 \tag{4}$$

where α_i = anisotropic thermal expansion matrix

T = temperature increase from a reference (stress-free) temperature. Stress resultants (N_i) and stress couples (M_i) can be defined as

$$[N_i, M_i] = \int_{-h/2}^{+h/2} \sigma_i \, [1,z] \, dz \tag{5}$$

Substituting Equations (3) and (5) into (6) renders:

$$\bar{N}_i = N_i + N_i^T = A_{ij}\epsilon_j^o + B_{ij}k_j \tag{6}$$

$$\bar{M}_i = M_i + M_i^T = B_{ij}\epsilon_j^o + D_{ij}k_j \tag{7}$$

where

$$[A_{ij}, B_{ij}, D_{ij}] = \int_{-h/2}^{+h/2} Q_{ij} \, [1,z,z^2] \, dz \tag{8}$$

$$[N_i^T, M_i^T] = \int_{-h/2}^{+h/2} Q_{ij} \, \alpha_j \, T \, [1,z] \, dz \tag{9}$$

The brackets above are symbolic rather than operational; the equality applies to the corresponding terms in the bracket. A_{ij} is the in-plane stretching stiffness; D_{ij} is the bending stiffness and B_{ij} is a coupling stiffness matrix due to nonsymmetric lamination.

If the ply illustrated in Figure 1 is used in the construction of a balanced and symmetrical 0/90 laminate and is mechanically tested, a bi-linear stress-strain curve is observed up to rupture. This initial slope of the 0/90 curve or moduli is the sum, through the thickness, of the plane-stress stiffness of each layer. As the laminate is deformed, each ply possesses the same in-plane strain, ϵ^o, and when the strain on the 90° layers reaches the strain level at which ply failure was experienced, the 90° layers crack (craze), shown in Figure 2. In a glass/resin system the occurrence of crazing is associated with audible noise similar to the creaking of a wooden mast, and a whitening of the material. The failure of the 90° layers in the laminate prevents the 90° layers from carrying their share of the load. This load is transferred to the unbroken layers, the 0° layers for our illustration, and results in a loss of laminate stiffness or modulus. Continual loading will ultimately produce a catastrophic failure of the laminate when the strain capability of the unbroken 0° layers is exceeded. For a 0/90 construction, employing the glass/epoxy material specified in Table 1, the ratio of the ultimate failure stress to the crazing stress, the knee in Figure 1 is 6.1. Experimental data and a theoretical stress-strain curve based upon the data of Table 1 are shown in Figure 3 for a $\pi/4$ (0°/± 45°/90°) glass/epoxy laminate. Note the

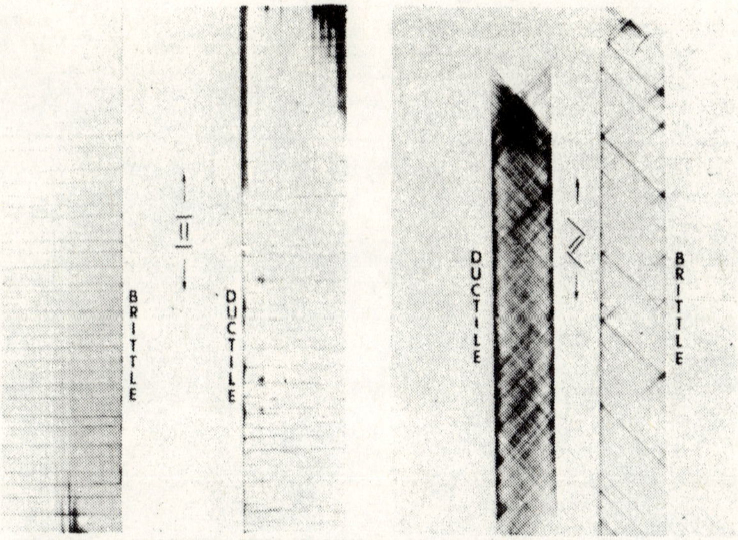

Figure 2. Crazing in glass-epoxy specimens for both brittle and ductile matrix systems, after Ishai and Lavengood [5].

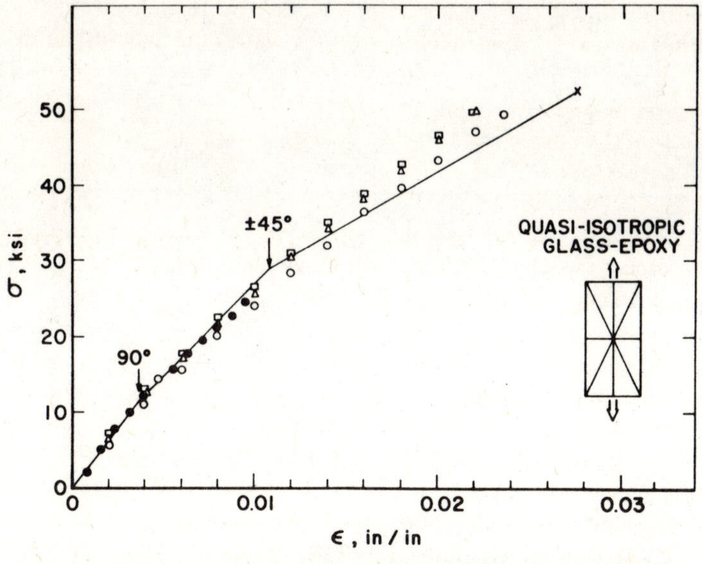

Figure 3. Comparison of theory and experiment for a glass-epoxy laminate.

Table 1. Unidirectional Lamina Design Properties as a Function of Material

Lamina Material (Density)	*Engineering Moduli, 10^6 psi*			*Poisson's Ratio*	*Maximum Strain, percent*			*Expansion, 10^{-6} in/in/°F*	
	E_{11}	E_{22}	G_{12}	ν_{12}	$\pm\epsilon_1$	$\pm\epsilon_2$	ϵ_6	α_1	α_2
Glass Epoxy (0.064 #/in³)	5.6	1.2	.6	0.26	2.75 1.58	.38 1.42	2.80	3.5	11.4
PRD 49-1 Epoxy (0.050 #/in³)	15	.83	.6	0.21	1.01 0.25	.25 .90	1.00	–0.6	–
High-Strength Graphite-Epoxy (0.054 #/in³)	21	1.7	.65	0.28	.72 .98	.45 .90	1.50	–0.4	16-18
High-Modulus Epoxy (0.058 #/in³)	27.6	1.1	.8	0.21	.45 .28	.30 2.60	.70	–0.55	–
Boron Epoxy (0.073 #/in³)	30	2.7	.65	0.21	.60 1.13	.35 1.13	1.95	2.5	13.0
Boron Aluminum (0.0975 #/in³)	33	21	7.0	0.23	.50 .55	.10 .10	.15	3.2	10.6
HM-HS Graphite Epoxy (0.052 #/in³)	46.0	1.0	.6	0.25	.16	–	–	–(0.5-.65)	(17-33)

change in stiffness as the 90° and then the 45° layers craze and the correspondence of the theoretical ultimate strength of 52,000 psi with the experimental results of 50,500 psi. The ratio of ultimate stress to 90° ply failure is 4.5. While the strain for transverse ply failure is constant from laminate to laminate, the stress required to craze the system, as well as final failure load, is a function of laminate geometry (Figure 4) because the construction of the laminate specifies the stiffness properties (crazing stress = stiffness × transverse ply failure strain). Therefore, lamination permits the engineer to tailor a fixed prepreg system to meet the conflicting stress/strain demands at different points in a structure. A further point, the crazing stress, or threshold, is generally at or below the creep fracture or fatigue limit for all classes of composites (for glass/epoxy the fatigue limit lies between 0.25-.30 of static ultimate strength). Boron and graphite are fatigue insensitive filaments, thus no fatigue damage is realized below first ply failure.

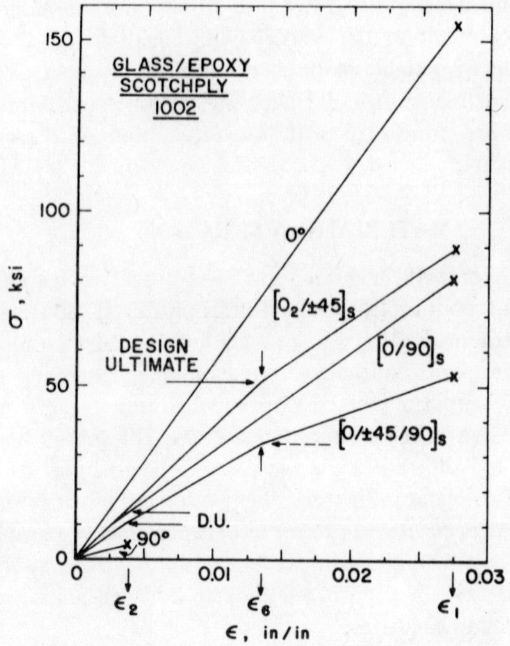

Figure 4. Influence of lamination upon the uniaxial tensile strength of a composite material.

Thus, the material properties of a laminate can be specified in terms of the ply engineering moduli; E_{11}, E_{22}, ν_{12} and G_{12}; the engineering strains to failure; ϵ_1, ϵ_2 and ϵ_6; and the thermal expansion coefficients; α_1 and α_2. Engineering properties of currently commercially available prepreg materials are tabulated in Table 1 for reference purposes.

DESIGN PHILOSOPHY

In order to evaluate high-modulus, fiber-composite technology, we must commit ourselves to a design concept. Do we design to the finial catastrophic failure of a laminate or to the first occurrence of mechanical damage [6] in the laminate? Because current research efforts indicate that the stress levels required for safe-life fatigue; for restricted environmental degradation on the composite filaments through a crazed matrix; and for controlled stiffness properties are in the vicinity of the stress levels predicted for first-ply failure, we have adopted this condition for illustration purposes.

Design ultimate loads shall result in a stress that does not exceed the *design ultimate stress* for the laminate used, where *design ultimate stress* is the maximum laminate stress attainable without rupture of any lamina.

Material stiffness and density also enter into an evaluation scheme. This density-stiffness sensitivity is a result of the eternal requirement of enhanced engineering property levels at reduced weight. Material stiffness allows control of the product shape and structural performance under buckling, flexural and vibrational loading conditions. Thus, we are concerned with the interaction of these variables in a material evaluation scheme.

MATERIALS EVALUATION

Emphasis upon high-strength, high-modulus fiber properties has characterized the composite development philosophy of the Fifties and early Sixties. This emphasis upon enhanced filament properties was strongly influenced by the belief that enhanced fiber properties were sufficient to guarantee engineering acceptance. In Figure 5, we have presented a plot of the modulus and strength in the filament direction for prepregged advanced composite materials. The strengths are the product of $E_{11} \cdot \epsilon_1$ of Table 1. Note that while the actual strengths may not be as high as some people report, these values represent engineering conservatism in that they are the *lower bound statistical limits on the ply properties.* The important point is that the strength is roughly constant, with the exception of the graphite fibers, with increasing filament stiffness. A typical relationship between graphite fiber stiffness and strength is illustrated in Figure 6.

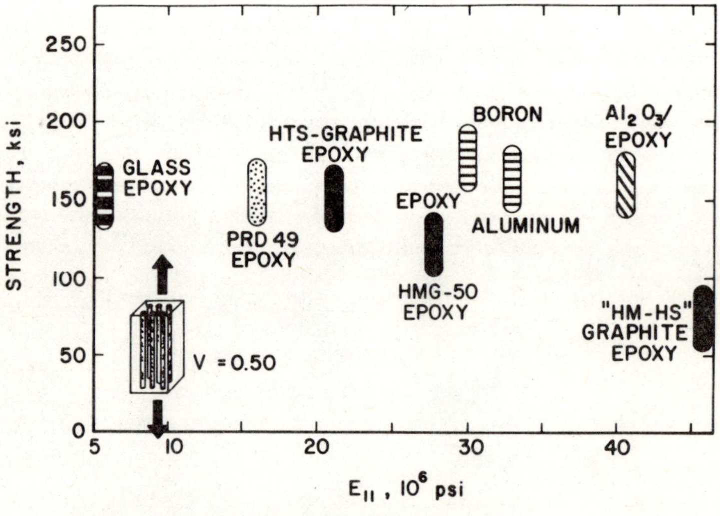

Figure 5. Comparison of longitudinal engineering strength against longitudinal modulus for typical prepreg materials.

Employing the data of Table 1 and the maximum strain criterion, uniaxial tensile properties were computed for two engineering laminates: a 0/90 system which

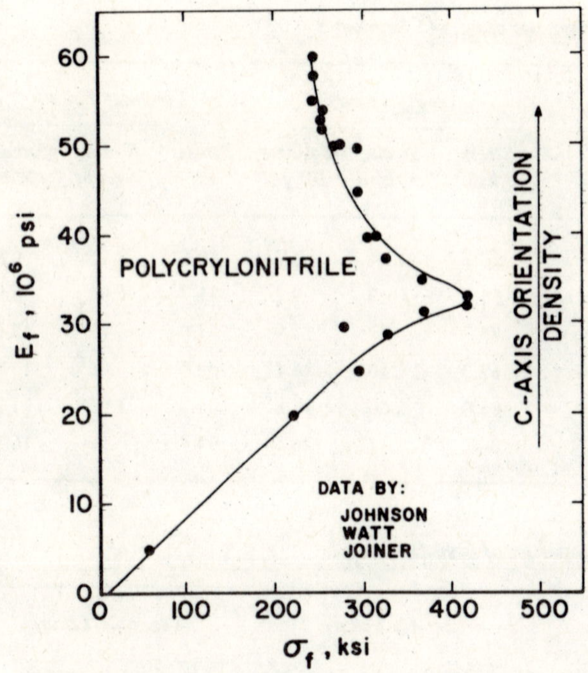

Figure 6. Illustration of the interrelationship between fiber modulus and strength for a typical graphite system.

possesses equal biaxial properties, and a $0_2/\pm 45$ system which has increased uniaxial strength properties and shear stiffness with reference to the 0/90 system. Both laminates possess similar initial stiffness and ultimate strengths, see Tables 2 and 3 and Figure 7. The *design ultimate load,* first ply failure, is the useful engineering capability of these systems. Increased design level of $0_2/\pm 45$ family over 0/90 family may be attributed to the increased strain capability of $\pm 45°$ layers, as compared to $90°$ layers. The relative ordering of design ultimate strength properties around the indicated dash lines is a direct reflection of transverse strain and modulus capability, ϵ_2 and E_{22} of Table 1. Both the DuPont PRD-49 fiber and the high modulus graphite fiber, "prepreg," have transverse properties which are partially attenuated by high filament anisotropy. The poor showing by the boron/aluminum system is the result of a low transverse strain at yield which offsets the large transverse stiffness advantage of the system. Note that the uniform ultimate strength capability exhibited in Figure 5 limits the engineering utilization of advanced filaments possessing stiffness properties in excess of boron. The boron/aluminum results also illustrate the limitation imposed by matrix properties. For example, a high modulus resin matrix for boron would yield a loss in composite properties if transverse strain capability is decreased. The dash line in Figure 7 roughly correlates with a transverse strain of 0.3%.

For illustration purposes, let us compare the composite properties illustrated in

Table 2. Tensile Properties of $0_2/\pm 45$ Laminates

	1st Ply Failure			Laminate Failure		
Material System	Stress, ksi	Strain, percent	Modulus, 10^6 psi	Stress, ksi	Strain, percent	Modulus, 10^6 psi
Glass/Epoxy	50.1	1.34	3.7	89.6	2.75	2.8
PRD 49-1/Epoxy	54.7	.60	9.2	87.7	1.01	8.0
HTS Graphite/Epoxy	85.7	.72	11.9	87.0	.83	12.0
H.M. Graphite/Epoxy	62.3	.40	15.4	68.6	.45	13.8
Boron/Epoxy	99.9	.60	16.6	105.3	1.05	1.2
Boron/Aluminum	30.2	.11	26.5	94.4	.50	16.5

Table 3. Tensile Properties of 0/90 Laminates

	1st Ply Failure			Laminate Failure		
Material System	Stress, ksi	Strain, percent	Modulus, 10^6 psi	Stress, ksi	Strain, percent	Modulus, 10^6 psi
Glass/Epoxy	13.0	.38	3.4	79.4	2.75	2.8
PRD 49-1/Epoxy	20.8	.60	8.4	81.9	1.01	8.0
HTS Graphite/Epoxy	51.2	.45	11.4	79.6	.72	10.5
H.M. Graphite/Epoxy	43.1	.30	14.4	63.8	.45	13.8
Boron/Epoxy	56.6	.35	16.4	94.8	.60	15.0
Boron/Aluminum	27.2	.10	27.0	93.5	.50	16.5

Figure 5 on a specific basis: property level divided by density. It is assumed in material comparison that laminate geometry and item geometry are fixed, and we wish to establish the "thickness" of the item required to satisfy the design load at a constant weight. For laminate strength, we would normalize composite properties as σ_i/ρ. First sizing of an aircraft item is performed on a strength basis. The second step in the sizing of a laminate would be on a stiffness basis for control of buckling and occasionally vibrational resonance stability. Vibration control at first resonance [7] would scale as $E_{11}/\rho^{3/2}$. Stiffness critical applications involves the bending rigidity of a laminate which varies as a cube of the thickness [8], or E_{11}/ρ^3. A plot of

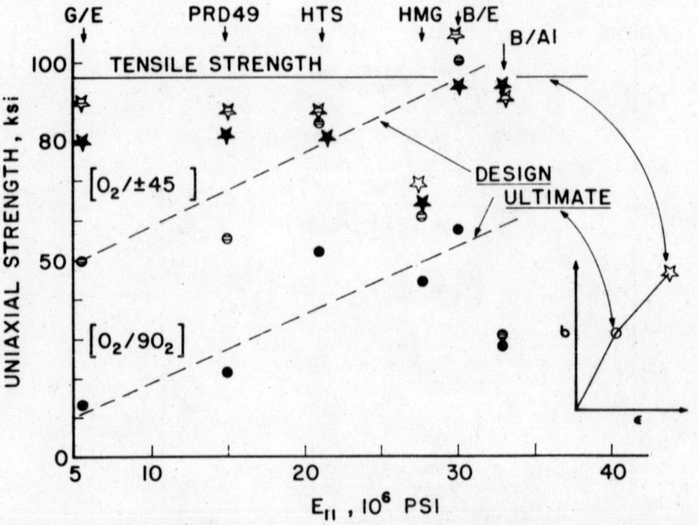

Figure 7. Interrelationship between design ultimate laminate capability and longitudinal prepreg stiffness; circles represent first ply failure; stars, laminate failure; solid symbols, 0/90 laminates; open symbols, $O_2/\pm 45$ laminates.

specific tensile strength against specific bending stiffness for the laminates considered in Figure 5 and Table 2 is shown in Figure 8. It is rather curious to observe that the graphite fiber which possesses a lower modulus is, in fact, the optimum material. The density difference between H.T.S. and HMG fiber is just sufficient to offset the modulus advantage of high modulus graphite fibers in stiffness critical applications. Also, observe that the density of the boron filament inflicts a penalty upon this composite. This density penalty may be sufficient to mitigate the compressive strength ($-\sigma_1 = E_x \cdot [-\epsilon_1]$ of Table 1) advantage of boron over graphite. A similar comment is pertinent to a sapphire-resin composite ($\rho \sim 0.110$ lbs/in^3). In this regard, the respectable properties of PRD-49 are lost in compression due to an inordinately low compressive failure strain $-\epsilon_1 \sim 0.0025$ or stress (−35,000 psi). It is this worker's opinion that highly anisotropic fibers possess poor compression properties for the analogous reason that increased anisotropy decreases the buckling load in an unidirectional lamina [7].

It is also interesting to note that the analysis presented here emphasizes operating strain requirements comparable to the engineering experiences in the metallic tradition. Traditional metallic design is based upon a 0.4 to 0.5% elastic strain capability, combined with a capacity for plastic deformation of 4-20% at high stress concentrations, (see Figure 9). A "safe-life" fatigue design results in limit strains of 0.2-0.3%. Accordingly, we would expect a lamina to possess a minimum strain

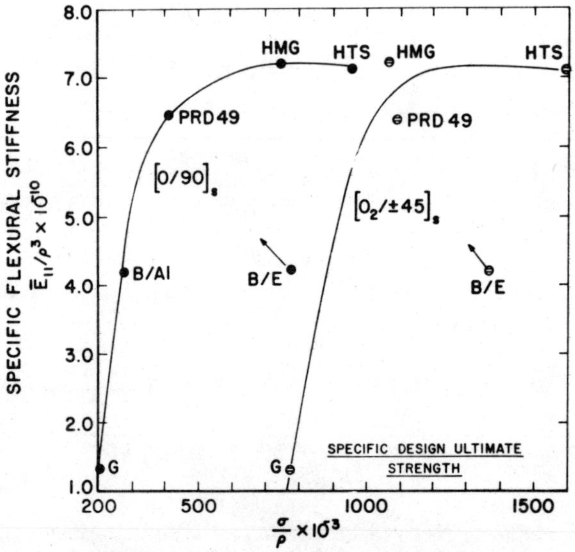

Figure 8. Specific flexural versus design ultimate properties for composite laminates in Figure 7.

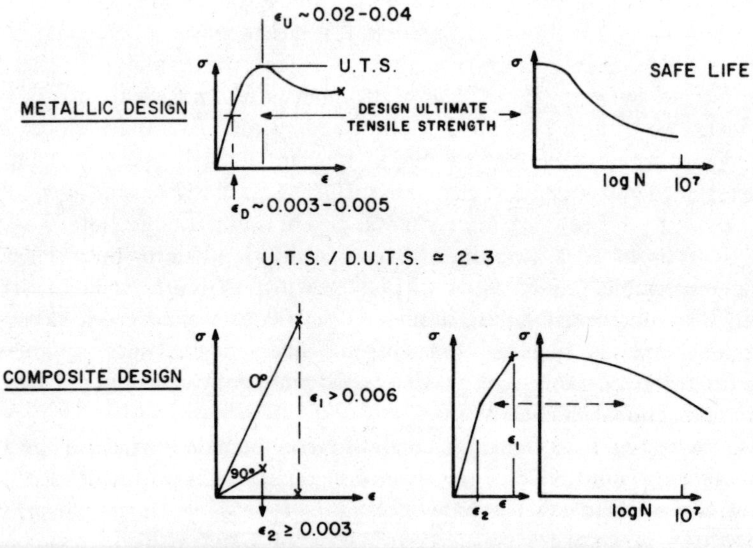

Figure 9. Comparison of metallic design requirements against composite design requirements.

allowable of this magnitude or:

$$\epsilon_2 \geq 0.003$$

The transverse lamina response is dominated by the matrix response, filament concentration, etc.; consequently, this allowable is a criteria to judge a specific matrix system performance in composite applications. Volumetric concentration of filaments is not a true design variable as balanced engineering and fabrication requirements fix this at roughly 50-60%. Longitudinal properties must exceed the transverse response as the longitudinal property provides the insurance or overload capability for the laminate. In addition, a balanced response in compression and tension is required for general engineering utilization or:

$$\epsilon_1 \geq \pm 0.006$$

Note that longitudinal tensile strain is essentially that of the reinforcement fibers. Experience also indicates that the shear strain should be comparable to/or greater than the longitudinal allowable:

$$\epsilon_6 \geq \epsilon_1$$

These criteria clearly demand a balanced materials evaluation program with emphasis placed upon quality control from fiber generation through prepregging to fabrication. The emphasis upon reproducible quality design allowables is the force favoring dry prepreg technology over the traditional wet filament winding technology.

Metallic matrices have not demonstrated the same performance capability available at ambient conditions currently available with resin matrix composites. Fabrication technology, limited by the chemical sensitivity of the reinforcement to the matrix at elevated temperatures, constitutes a formidable obstacle to general engineering acceptance. Extensive effort is being expended to develop metal matrix systems to the same acceptable performance levels.

DIMENSIONAL STABILITY

Composite materials find application where high specific properties, chemical inertness, low maintenance, dielectric capability and dimensional stability are required. Problems related to dimensional stability are particularly well suited for composite application, as currently available filaments possess a wide range of technically useful properties. An inspection of Table 1 will indicate that the PRD 49-1 and the graphite filament family yield negative thermal expansion coefficients combined with high axial stiffness along the fiber axes. Combining fibers of this type within a matrix, we may generate a stiff, strong laminated solid with exceedingly small, if not zero in-plane, expansion. This result is achieved when the fibers are of sufficient stiffness to dominate the axial response of the laminate.

Referring to Equation (4) we may express the thermal expansion properties of lamina in terms of fiber and matrix by Schapery [9] as:

$$\alpha_1 = \frac{E_m \alpha_m V_m + E_f \alpha_f V_f}{E_m V_m + E_f V_f} \tag{10}$$

and

$$\begin{aligned} \alpha_2 = {} & (1 + V_m)\alpha_m V_m + (1 + V_f)\alpha_f V_f \\ & - \alpha_1 (V_f V_f + V_m V_m) \end{aligned} \tag{11}$$

For anisotropic filaments, axial and transverse fiber properties are employed in the respective calculation for α_1 and α_2. Once the expansion coefficients of a single layer of fiber reinforced material have been computed from constitutive material properties, the response of a laminate can be determined through the use of lamination theory [3].

$$\begin{aligned} \bar{\alpha}_1 &= \frac{A_{22}R_1 - A_{12}R_2}{A_{11}A_{22} - A_{12}^2} \\ \bar{\alpha}_2 &= \frac{A_{11}R_2 - A_{12}R_1}{A_{11}A_{22} - A_{12}^2} \end{aligned} \tag{12}$$

where

$$R_1 = J_1 h + J_2 \sum_{n=1}^{n} h_n \cos 2\theta_n$$

$$R_2 = J_1 h - J_2 \sum_{h=1}^{n} h_n \cos 2\theta_n$$

$$J_1 = (U_1 + U_4)W_1 + 2U_2 W_2$$

$$J_2 = U_2 W_1 + 2W_2(U_1 + 2U_3 - U_4)$$

$$W_1 = (\alpha_1 + \alpha_2)/2$$

$$W_2 = (\alpha_1 - \alpha_2)/4$$

The laminate calculation is somewhat surprising in that it suggests that at certain angles, the laminate will contract in the longitudinal direction when heated. This prediction has been confirmed by Halpin and Pagano [3, 10] and Fahmy and Ragai [11] and these results are illustrated in Figure 10 for both positive and negative α_1. In Figure 11 is shown photographs of a model system undergoing swelling expansions due to absorbed liquids. The angle ply $\pm\theta$ in (b) is shown before an expansion. Note

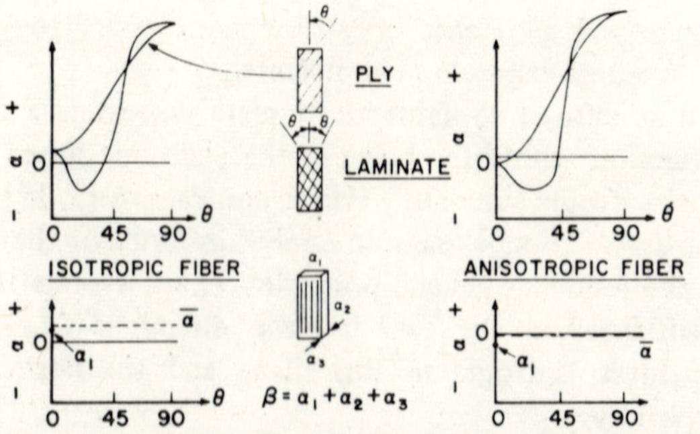

ISOTROPIC EXPANSION FOR LAMINATES

$$\bar{\alpha} = \alpha_1 = \alpha_2 = W_1 + \frac{2(E_{11}-E_{22})}{E_{11}+(1+2\nu_{12})E_{22}} W_2 \qquad \text{AS } E_{11}/E_{22} \rightarrow \infty$$

$$W_1 = 1/2\,(\alpha_1^L + \alpha_2^L) \qquad \bar{\alpha} \rightarrow W_1 + 2W_2 \rightarrow \alpha_1^L$$

$$W = 1/4\,(\alpha_1^L + \alpha_2^L) \qquad \text{IF } E_f/E_m > 100,\ \alpha_f < 0$$

$$\bar{\alpha} \rightarrow 0$$

Figure 10. Influence of lamination upon thermal expansion properties.

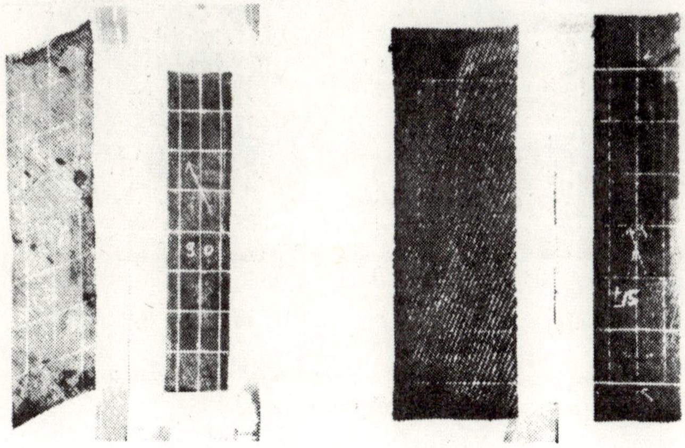

(a) (b)

Figure 11. Off-axis and angle-ply laminate expansion experiment.

the unusual negative longitudinal contraction and the correspondingly large transverse expansion. In (a) of the figure is shown the expansion response of a unidirectional off-axis ply. Compare the large longitudinal and transverse expansion as compared with the angle ply. Also, note the shear response due to the angular rotation. Results of this type have obvious technological significance as they specify the material parameters and geometric construction necessary to yield structural elements with zero or minimal expansion coefficients.

For example, if it is desired to fabricate a plate possessing a zero expansion coefficient in any direction within the plane of the plate one solves the equivalent problems of the "quasi-isotropic laminate." Halpin and Pagano [3, 12] have obtained a solution for the "quasi-isotropic expansion strain" in terms of the ply properties. This solution asserts the following general principle: for any symmetric laminate ($B_{ij} = 0$) with equal stiffnesses in the two in-plane directions ($A_{11} = A_{22}$), the expansional strain field is isotropic in this plane and the normal strain in all directions is simply given by:

$$\bar{\alpha} = \alpha_1 = \alpha_2 = W_1 + \frac{2W_2(E_{11} - E_{22})}{E_{11} + (1 + 2\nu_{12})E_{22}} \tag{13}$$

The experimental proof is shown in Figure 12 for the same material shown in the earlier figure. Note the very small dimensional changes of the expanded item (b). This very small expansion strain is in accord with theory which states that as $E_{11}/E_{22} \rightarrow \infty$, the isotropic expansion strain becomes the same in the longitudinal

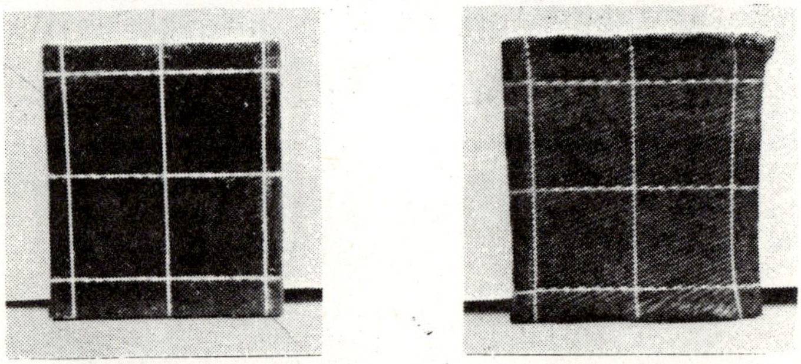

(a) (b)

Figure 12. Experimental verification of minimization of expansion through lamination.

unidirectional ply $e \rightarrow e_1^L$. Thus, for very highly anisotropic composites (high ratio of longitudinal to transverse properties), the isotropic strain approaches the expansion strain of a unidirectional ply parallel to the filaments. To achieve a zero coefficient of expansion one should have:

$$E_f/E_m > 100 \tag{14}$$

$$\alpha_f < 0 \tag{15}$$

Zero, isotropic expansion coefficients have been technically achieved in this manner. Specialized, commercial applications demanding dimensional stability will generate a market potential for an anisotropic filament reinforcement, in addition to the market for primary and secondary structural applications.

RELIABILITY AND SAFETY OF COMPOSITE STRUCTURES

The structural reliability of an item is dependent on the likelihood of a static or a fatigue failure. The specific goal of a reliability analysis is to determine the expectation of first damage due to static or fatigue failure processes in structures or specific structural detail. This goal is confronted with the confounding elements of material variability, fabrication, quality control, design errors, assembly variables, statistical load environment and physical environments as illustrated in Figure 13. Structural safety in aircraft design is implemented with two sets of safety factors and a design philosophy.

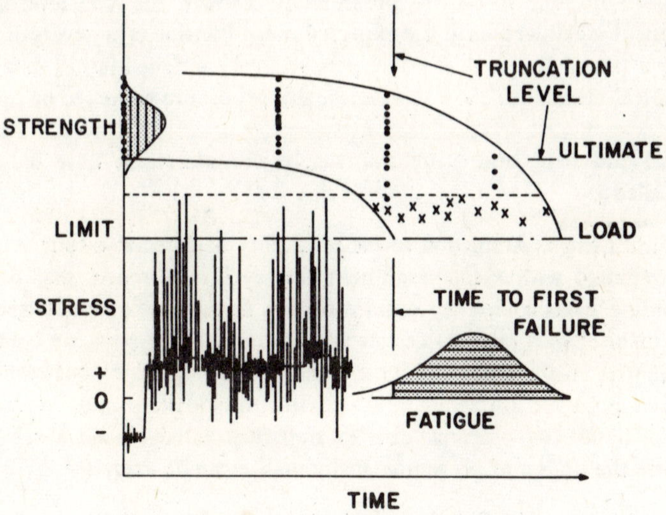

Figure 13. Illustration of relationship between static fracture, engineering load history and fatigue.

Ultimate Factor of Safety: A margin of strength capacity is desired in excess of the major portion of the load spectrum (generally the most severe gust loads [13]) to insure against static fracture of the structures due to the above indeterminate variables. This operating load level is known as limit load, see Figure 13, and represents a safe operating load condition. For historical reasons the ratio of the limit load to the ultimate design load is 1.5. It is generally recognized that infrequent peak asymmetric maneuver loads exceed the limit load by a factor of 1.2. Due to the low static strength variability inherent in metallic structures, these overload conditions are acceptable. The loads at fatigue critical locations are often reduced by an additional safety factor of 1.15—1.30 on the limit load condition (for example, (1.5) times (1.3) gives 1.95 on the Design Ultimate Load).

Indefinitely Long Life: Based upon the assumption that no perceptible fatigue failure occurs if peak stresses are less than some value called a fatigue limit (the plateau in the *S-N* diagram illustrated in Figure 9).

This approach assumes the existence of a fatigue limit exhibited by most steels, but not always exhibited by aluminum alloys, and a structural weight inherent in a design limited by extremely low stresses under infrequent high loads. These problems led to the concept of:

1. *Safe Life Design.* Admission of a *finite* but usefully long trouble free life time after which the structural item is to be retired or repaired. Implementation involves the use of the scatter factor, probabilistic load history characterization, relatively low fatigue allowables and rigorous quality control to minimize materials variation.

2. *Scatter Factor.* Due to the fact that fatigue lifetimes exhibit statistical scatter and the mean lifetime is only approximately known, the load level in a fatigue critical element is reduced until the expected mean lifetime is in excess of the desired lifetime by a factor of 2 to 4. If the dispersion in the fatigue lifetime data is small, as it is in classical aluminum alloy structures, this procedure proves to be operationally useful.

Technical problems and structural weight considerations have led to a third approach called:

3. *Fail-Safe Design.* Admission to the possibility of detectable fatigue crack development, combined with the commitment that such occurrences shall be prevented from becoming catastrophic by nondestructive damage detection, inspection, and preventive maintenance prior to catastrophic failure. These steps are combined with design such that small, detectable cracks permit safe operation between inspection periods. The successful implementation of this design philosophy presupposes the definition of a safe-life precrack time (a statistical variable) and a safe inspection period before the option of preventive maintenance can be exercised.

The adoption of a *Fail-Safe Design* philosophy is in effect a tacit admission that advanced structural design load levels (combined with increased yield stress metals) are at such a point that optimized lightweight metallic structural systems are not dominated by static fracture, but by the probability of fatigue cracking. These trends which have produced an increase in the occurrence of premature service fatigue failures [14], substantially increased maintenance expenditures, and a vocal discontent where driven by the dual need for weight and cost savings. See Table 4 for typical fatigue data. The demonstrable technical need for materials compatible with advanced structural requirements is the driving force behind the developing composite technology as summarized in Figure 9.

The fatigue characteristics of laminated composites are such that technically acceptable specific load levels are accessible on the basis that no fatigue failures in composite laminates are acceptable during the vehicle lifetime. This situation results from the fatigue insensitivity of the graphite and boron filaments combined with the

Table 4. Comparative Fatigue[1]

Material	*Maximum Stress per Cycle at 10^7 cycles (ksi)*
Boron or Graphite	
0	110
Boron or Graphite	
Epoxy	
$0_3/45$	85
$0_2/90_2$	58
D6AC Steel	
220-240HT	38
6-4 Titanium	35
2024T 851 Aluminum	13

[1] Constant amplitude fatigue
$k_t = 5$, $R=0.1$ room temperature

fact that laminated composite solids are internally redundant structures. Consequently, simple monolithic crack growth characteristics, such as those exhibited in metallic solids, are not easily observed in laminated composites.

This internal redundancy is balanced by the occurrence of a "static-notch effect" in laminated materials but this notch effect, in contrast with metallic response, disappears under fatigue conditions [15]. Thus, a safe-fatigue life offers protection against static failures. These behavioral characteristics are summarized by the term "quasi-brittle" and the following statements:

1. Static strength distributions for laminates controlled by filament failures are dictated by static strength distribution of the 0° prepreg direction

$$P(\sigma) = \exp-[\sigma/\hat{\sigma}]^{\alpha} \tag{16}$$

where $P(\sigma)$ is probability of N specimens out of m initial specimens surviving a static stress, σ, α is a shape parameter, and, $\hat{\sigma}$ is called a location parameter where the form of the equation is assumed to be Weibull.

2. Effect of notches is to reduce the static strength without a change in the shape of the distribution; a dependence of $\hat{\sigma}$ upon notch effects but not α; see Table 5 and [15, 16].

3. Although fatigue insensitive, laminated composites will accumulate damage and ultimately fail due to this damage accumulation. In lieu of a refined nondestructive testing capability, we [17] may study the effects of this damage accumulation by assuming that the rate of damage C is

Table 5. Comparison of Weibull Parameters for HTS-Epoxy Laminates[1]

Angular Content θ	Shape Parameter α	Location Parameter $\hat{\sigma}$ (ksi)
0	10.11	151.8
0/90	10.91	76.5
0/±45	10.8	66.9
0/±45/90	11.46	49.4
0/±45/90		
with round hole	11.5	39.2
with a slit	10.8	39.8
90	7.54	5.3

[1] Properties based upon early commercial prepreg. Laboratory and other commercial prepreg yields α values of 16-20 in magnitude.

$$\frac{dc}{dt} \simeq M C^r; \quad r \geq 1.0 \tag{17}$$

where M is a history dependent quantity and r is a material constant. In addition, assume that laminate failure occurs when

$$\sigma \sqrt{c} = K \tag{18}$$

where K is an apparent toughness or work parameter.

Employing Equations (16), (17) and (18) we can show that the laminate will possess a fatigue lifetime curve of the form:

$$\hat{t}_b \, \sigma_b^{2r} = \text{material constant} = B; \tag{19}$$

and a continuous change in static strength of the ith specimen expressed as:

$$t = \frac{1}{(r-1)M} \cdot \frac{1}{K^{2(r-1)}} \left[\sigma_i(0)^{2(r-1)} - \sigma_i(t)^{2(r-1)} \right] \tag{20}$$

when the strength $\sigma_i(t)$ reduces to the peak stress in the load history illustrated in Figure 13, the specimen breaks producing a fatigue failure, t. If the initial static strength distribution, Equation (16), is assumed to represent a distribution in the damage parameter C, Equation (18), and each damage site develops according to Equation (17) that the fatigue lifetime distribution will be given as:

$$P(t) = \exp-[t/\hat{t}]^{\frac{\alpha}{2(r-1)}}$$

The material term r is obtained from Equation (19) or from the derivative of (20):

$$\frac{d\sigma_i(t)}{dt} = \left[\frac{1}{2} MK^{2(r-1)}\right] \sigma_i(t)^{-(2r-3)} \tag{22}$$

which is an indirect measure of Equation (17). Preliminary material constants α and r are presented in Table (6).

Table 6. Comparison of Damage Rate Parameters

	r	α (Static)	$\alpha 2/(r-1)$ (Fatigue)
E-Glass/Epoxy	5.64	12.3	∿ 1.32
Boron/Epoxy	9-11	18-23	0.9-1.1
Graphite/Epoxy	8.5-10	17-19*	0.9-1.1
Aluminum	4-6	30-45	4-5

*High-quality prepreg

The results of these operations are illustrated in Figure 13 where we observe that the static and fatigue lifetime shape parameters are causally related to each other. In addition, note that the distribution of residual static strengths broadens in time and is truncated in the vicinity of the maximum load peaks in the variable history. Thus, the weakest numbers of the static strength population produces the first fatigue failures. The specification or location of the weakest member in the population, for a given load history, specifies the magnitude of the expected safe-life of the item. Truncation of the static strength distribution results in a truncation in the lifetime distribution function.

For example, consider the proof test experiment demonstrated in [16] for the 0/±45/90° glass epoxy laminate treated in Figure 3. Following the procedure and results of Figure 2 in [16], the stated truncated distribution is defined as:

$$P_p(\sigma) = \frac{P(\sigma)}{P(\sigma_p)} \tag{23}$$

where σ_p is the stress level below which no specimen will fail after truncation. For this illustration $P(\sigma_p)$ was equal to 0.812 (24 out of 30 specimens remained after the truncation). These specimens were then fatigued with a Gaussian load history until fatigue failure occurred. The truncated lifetime [17] distribution is expressed as:

$$P_p(t) = \frac{P(t)}{P(\sigma_p)} = \frac{\exp-[t/\hat{t}]^{\alpha/2(r-1)}}{\exp-[\sigma_p/\hat{\sigma}]^{\alpha}} \tag{24}$$

and the results are illustrated in Figure 14. Note that the requirement for a useful proof test is that the surviving specimens are not noticeably altered in their fracture properties. In this illustration we have an extreme case in that we have truncated the static specimens at 47 ksi, while the design ultimate for this laminate construction is only 10–11 ksi. Obviously, serious effort is required in this research area paralleling the work of Tiffany [18].

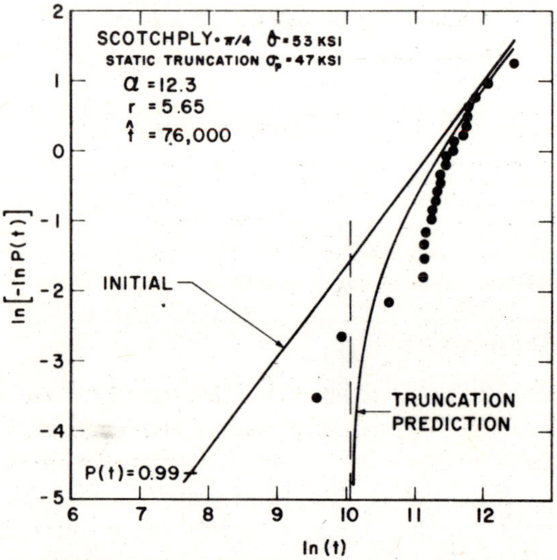

Figure 14. Truncation of fatigue lifetimes by a static-proof test [14].

The ranking of structural items based upon static response characteristics also contains the implicit assumption that the delayed time dependent fracture or fatigue characteristics are an extension of static fracture modes. In fact, this supposition is not applicable to metallic structures joined with mechanical fasteners [19]. It is often observed, in metallic structures [14], that structural fatigue is initiated by fretting around the fasteners and accelerated by galvanic corrosion induced in the structural detail by the fasteners. Primary organic adhesive bonding is currently being introduced into airframe construction in an effort to minimize these effects which contribute to the indeterminate [19] nature of structural safety in metallic structures. Composite structural technology is also seeking to exploit the potential advantages of adhesive joining technology. Quality control in adhesive bonding is being developed and the techniques outlined above will play a serious role in this evaluation as conventional, nondestructive techniques are not sufficient for the job. Well made adhesive joints are reliable, exhibit predictable scatter properties and yield fatigue failure modes which are related to static failure modes. Thus, the results presented here [17], by Halpin, Kopf and Goldberg [16], by Fruedenthal [19], and

Tiffany [18] will provide the analytical bases for a composite structural technology which exhibits a significantly higher degree of reliability than its equivalent metallic counterpart.

PROJECTIONS AND CONCLUSIONS

It is this worker's personal opinion that the (civilian as well as aircraft structures) general use of advanced composites will center around a filamentary prepreg composite which exhibits a longitudinal composite strain capability at failure of at least ±0.007, a transverse ply strain capability of at least 0.003, a shear strain capability equal to or greater than the longitudinal capability, fatigue insensitivity (75-80% of ultimate at 10^6 cycles) in a filament direction and a small but negative thermal expansion coefficient parallel to the filament axis. The technical preference for a high-strength, high-strain fiber, an acceptable ratio of high strength to modulus, is augmented by economic considerations. Implicit in this personal opinion is the argument that over the next decade the unsubsidized commercial market will only sustain one high-modulus, *general purpose* prepreg in addition to the existing E-glass/epoxy system. The economic scale of a materials system is not only constrained by development-production costs and market volume, but by the time and costs required in the development of a behavioral and engineering data base necessary for utilization of a materials system. At this specific time carbonization graphitization technology (both polymeric fiber and pitch precursors) yielding intermediate modulus high-strength fibers is attracting considerable serious interest. Such a system possesses a wide range of product application which cannot be duplicated by existing metallic technology. Augmenting this material will be a series of other composite systems which fulfill specific needs in modern technology. Future developments may produce a high modulus ribbon or flake composite which would reduce the fabrication requirements for complex laminate geometry and a matrix system with increased thermal properties and decreased moisture absorption. The ultimate nonmetallic construction material may not be a heterogeneous material, but a homogeneous polymeric material highly oriented (on the molecular scale) in at least two in-plane directions and yielding mechanical properties comparable to that of high-strength, graphite-epoxy systems discussed here.

The insertion of composite solids into engineering design and structures will alter structural design philosophy. Obviously structural design geometry must reflect some changes in material characteristics, but more importantly, composite technology complemented by advanced structural reliability concepts will emphasize the attainment of safe, reliable, predictable structures. Preliminary evidence indicates that some necessary aspects of structural reliability begins at the materials development level. This effort will be supported by increased emphasis upon education and the further development of a reproducible, high-production-volume material.

REFERENCES

1. A. M. Lovelace and S. W. Tsai, "Composites Enter the Mainstream of Aerospace Vehicle

Design," *Aero and Astro*, Vol. 8 (1970), p. 54.
2. G. P. Peterson and S. W. Tsai, "Status of Development of Advanced Composites," Proceedings 4th Symposium on Comp. Mat.; Tokyo, Japan, August (1971).
3. J. E. Ashton, J. C. Halpin, and P. H. Petit, *Primer on Composite Materials: Analysis*, Technomic Pub. Co., (1969).
4. S. W. Tsai, J. C. Halpin, and N. J. Pagano, *Composite Materials Workshop*, Technomic Pub. Co., (1968).
5. O. Ishai and R. E. Lavengood, "The Mechanical Performance of Bidirectional Fiber-Glass Polymeric Composites," *Israel Journal of Technology*, Vol. 8 (1970), p. 101.
6. C. W. Rogers, "Structural Design with Composites," *Fundamental Aspects of Fiber Reinforced Plastic Composites*, edited by R. T. Schwartz and H. S. Schwartz, Interscience, N. Y. (1968).
7. J. E. Ashton and J. M. Whitney, *Theory of Laminated Plates*, Technomic Pub. Co., (1971).
8. J. M. Whitney, "Effects of Constituent Material Properties on the Stability of Fiber Reinforced Composite Plates," Ref. 6 above.
9. R. A. Schapery, "Thermal Expansion Coefficients of Composite Materials Based on Energy Principles," *J. Comp. Mat.*, Vol. 2 (1968), p. 380.
10. J. C. Halpin and N. J. Pagano, "Consequences of Environmentally Induced Dilation in Solids," *Proc. 6th Ann. Meeting, Society of Engineering Science*, Springer (1969).
11. A. A. Fahmy and A. N. Ragi, "Thermal Expansion of Graphite-Epoxy Composites," *J. Appl. Phys.*, Vol. 41 (1970), p. 5112. .
12. J. C. Halpin and N. J. Pagano, "The Laminate Approximations for Randomly Oriented Fibrous Composites," *J. Comp. Mat.*, Vol. 3 (1969), p. 720.
13. F. R. Shanley, "Historical Note on the 1.5 Factor of Safety for Aircraft Structures," *J. Aero. Sci*, (1962), p. 243.
14. H. J. Grover, *Fatigue of Aircraft Structures*, Navair 01-1A-13 (1966).
15. M. E. Waddoups, J. R. Eisenmann and B. E. Kaminski, "Macroscopic Fracture Mechanics of Advanced Composite Materials," *J. Comp. Mat.*, Vol. 5 (1971), p. 446.
16. J. C. Halpin, J. R. Kopf and W. Goldberg, "Time Dependent Static Strength and Reliability for Composites," *J. Comp. Mat.*, Vol. 4 (1970), p. 462.
17. J. C. Halpin, T. Johnson, and K. Jerina, unpublished results.
18. C. F. Tiffany, "On the Prevention of Delayed Time Failures of Aerospace Pressure Vessels," *J. Franklin Inst.*, Vol. 290 (1970), p. 567.
19. A. M. Freudenthal, "The Material Aspects of Reliability," *Structural Fatigue in Aircraft*, ASTM STP 404, 1966.

Tensile Properties of PRD-49 Fiber in Epoxy Matrix*

T. T. CHIAO AND R. L. MOORE, *Lawrence Livermore Laboratory, University of California, Livermore, California 94550*

(Received September 9, 1972)

Recently a high-modulus and high-strength organic fiber, identified as PRD-49-III by the manufacturer, became commercially available. To assess the fiber performance and its potential as a reinforcement for fiber composites, we have determined the tensile properties of the fiber in an epoxy matrix. During this study we made over 6000 fiber/epoxy strands using the filament winding process; these strands were representative samples taken from over 60 lb of single-end fiber. We were interested in the fiber uniformity, its strength distribution at room and LN2 temperature, stress-strain characteristics, and the strain rate effect on the fiber strength.

EXPERIMENTAL

Materials:

A popular epoxy, Union Carbide ERL 2258/ZZL 0820 (100/30), was used for our study. The cure schedule for all specimens was constant: gel at 200°F for 3 hr, then cure at 325°F for 2 hr. The tensile properties of the pure cast resin are reported separately [1].

The single-end (nominal 380 denier with 280 filaments) preproduction fiber was supplied without a finish, but with a producer's twist of about one turn per foot. The chemical structure of the fiber was suspected to be poly-p-benzamide polymer [2].

Specimen Preparation and Test:

A spool of fiber was first dried in a vacuum oven for a minimum of 24 hr at 180°F; approximately one layer was stripped from the outside and discarded. Then the fiber was epoxy impregnated and wound into strand specimens using a vacuum filament winding process [3]. One hundred specimens were made from each of the

*This work was performed under the joint auspices of the National Aeronautics and Space Administration – Lewis Research Center (Contract No. C-13980-C) and the Atomic Energy Commission.

61 fiber spools. From a random spool several hundred more specimens were made to study the effects of strain rate and LN_2 temperature on fiber tensile strength.

The linear weight of the fiber on each spool was based on the average weight of three 100-in. bare fiber specimens taken from the spool prior to test specimen fabrication. Five fiber/epoxy strands were selected at random from each of the first 24 spools for ultimate tensile stress determination only. Ten specimens were taken from each of the remaining 37 spools, for stress, strain, and modulus determination. In every case, one random fiber/epoxy strand from each spool was used to determine the fiber content.

Specimens of 10-in. gage length were tested at 0.5 in./min crosshead speed on a tensile test machine. The clamp arrangement was improved by bonding the specimens in the mechanical clamps with a room-temperature epoxy adhesive instead of the cushioning materials used in our previous work [4].

For modulus and strain measurements, we calibrated the compliance of our tensile machine by using various gage length specimens [5]. The modulus data were further confirmed using accurate strain-measuring linear variable differential transformers. The transformer cores were attached directly to the test specimen [6].

The study of strain rate effects on the fiber strength of the strands was based on the procedure reported previously [7]. The effect of LN_2 temperature on the fiber strength was determined by immersing the test specimens directly in LN_2.

RESULTS AND DISCUSSION

Since hundreds of tests were involved, a tabular data presentation was not practical. Instead, we made extensive use of probability plots. For easy comparison, the abscissa was arbitrarily normalized to the median value in each case.

The fiber volume content of the strand specimens varied more than we expected (11.0%). Figure 1 shows the scatter. Its effect on the fiber strength in a strand was not determined.

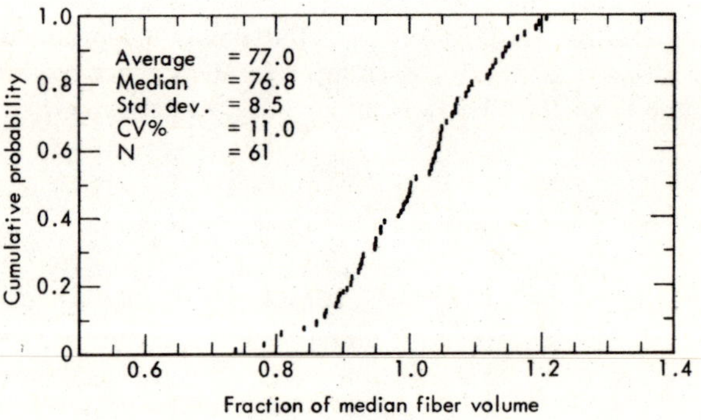

***Figure 1.** Distribution of volumetric fiber content of the test specimens – organic fiber/epoxy strands.*

The bare fiber weight per linear length (the textile term, denier, is g/9000 m) is an indirect measurement of the filament diameter variation. This data with the fiber density of 1.45 g/cc at room temperature allows a check of the cross-sectional area variation of the strands: average = 4.562×10^{-5} in^2; coefficient of variation = 2.1%; and the number of specimens = 187. It is clear that the fiber is quite uniform. It appears that this average area can be used as a constant to represent the single-end strand for engineering calculations. Photomicrographs of the strands (Figure 2) further confirm the fiber uniformity.

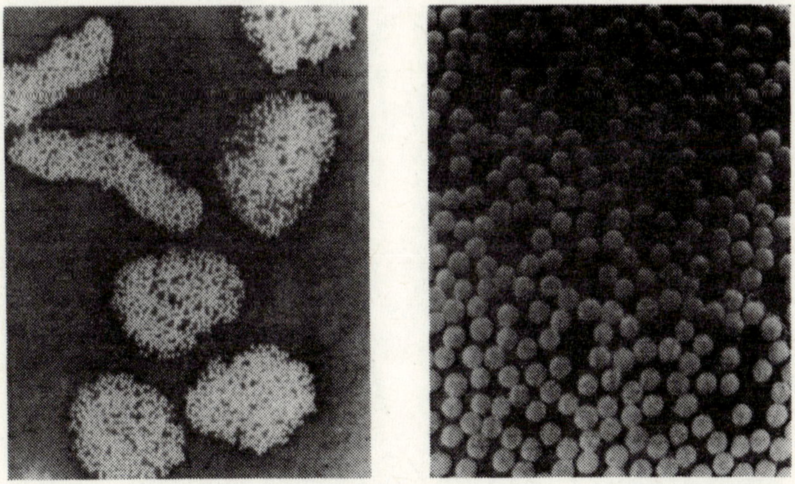

***Figure 2.** Photomicrograph of the cross section of the organic fiber/epoxy strands (top, 100×; bottom 500×).*

The fiber strength distribution based on the average cross-sectional area of individual spools, is shown in Figure 3. There are a few low values. The scatter, however, is reasonably narrow. The output of the tensile tester was recorded on magnetic tape for computer processing and plotting of the stress-strain data for the majority of the specimens. These results are summarized in Figure 4. The strain values were corrected based on a constant compliance of 8.0×10^{-1} in./lb. Modulus calculations were based on the stress at 1% strain.

The strain rate effect on the fiber strength is of interest to many researchers. Table 1 shows such data based on specimens from only one spool of fiber. The rate effect, if any, is minimal. Our recent test data on pressure vessels wound with the same material system confirm the strand data.

We also checked the effect of LN_2 temperature on the fiber strength of strands. Using specimens from the same spool of fiber, at 70°F the average fiber strength from 53 specimens is 505 ± 3.4 ksi with a coefficient of variation (*cv*) of 2.5%; at −320°F the average fiber strength from 27 specimens is 486 ± 9.9 ksi with a *cv* of 5.2%. Thus, the fiber strength reduction at LN_2 temperature is not significant, considering the data scatter at the low temperature.

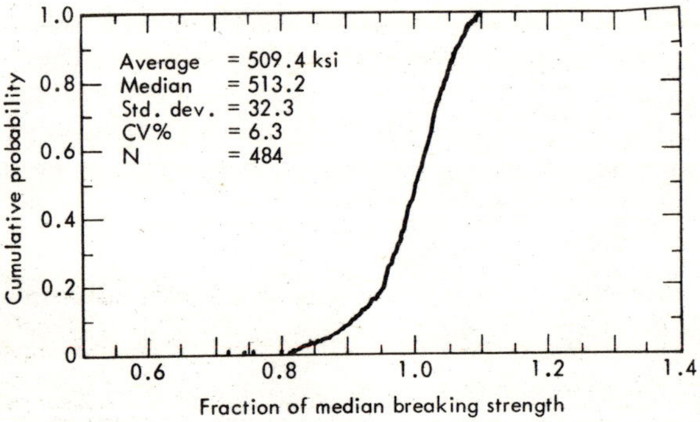

Figure 3. *Fiber strength distribution of the organic fiber/epoxy strands.*

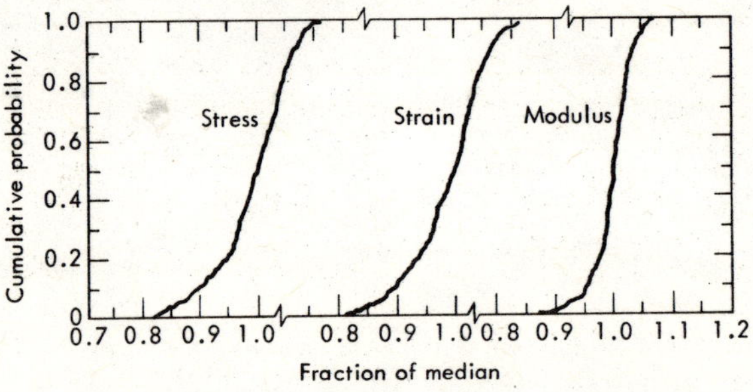

	Stress	Strain	Modulus
Average =	506.3 ksi	2.40%	19.97×10^6 psi
Median =	510.9	2.42	20.02
Std dev =	33.1	0.16	0.65
CV %	6.5	6.7	3.3
No. of specimen =	363	363	363

Figure 4. *Tensile properties from 37 spools of the organic fiber.*

CONCLUSIONS

The typical strength of the PRD-49-III fiber in epoxy matrix is 509 ksi at 70°F. The fiber uniformity in terms of filament diameter, denier, and cross-sectional area is quite acceptable as an engineering material. Both LN_2 temperature and strain rate have little effect on the fiber strength.

Considering its high modulus, high strength, and low density, this fiber is at present the most attractive reinforcement for fiber composites in tensile-critical applications where shear stress is minimal.

Table 1. Strain Rate Effect on the Ultimate Tensile Stress of the Organic Fiber/Epoxy Strands

Strain rate[a] *(in./in./min)*	*Avg. tensile strength*[b] *(ksi)*	*95% conf. limits*	*Std. dev. (ksi)*	*Coeff. of var. (%)*	*No. of specimens*
5	468	±12	9.6	2.1	5
2	463	±24	19.6	4.2	5
1	507	±26	20.7	4.1	5
0.5	494	±22	17.9	3.6	5
0.1	497	±27	21.6	4.4	5
0.05[c]	499	±12	9.9	2.0	5
0.01	501	± 4	3.1	0.6	5
0.005	490	±30	23.8	4.9	5
0.001	471	±19	17.6	3.7	6
0.0005	476	±16	12.8	2.7	5

[a]Crosshead speed divided by 10-in. gage length.

[b]All strands from one spool, the average cross-section area of 4.364×10^{-5} in.^{2} was used to calculate stress.

[c]Rate normally used for fiber strength determination.

ACKNOWLEDGMENT

We wish to thank W. D. Zinke of Sandia Laboratories, Livermore, California, for lending us the LN_2 test fixture used in our study.

REFERENCES

1. T. T. Chiao and L. P. Althouse, "Characterization of an Epoxy System for Filament Winding," Proceedings of the 4th National SAMPE Technical Conference, Oct. 17-19, 1972, Palo Alto, California.
2. T. T. Chiao and R. L. Moore, "Organic Fiber/Epoxy Composites," Composites, Jan/Feb. (1973).
3. T. T. Chiao and R. L. Moore, "Strength of S-Glass Fiber," *SAMPE Quarterly*, Vol. 3, April (1972), No. 3.
4. T. T. Chiao and R. L. Moore, "A Tensile Test Method for Fibers," *J. Composite Materials*, Vol. 4 (1970), p. 118.
5. W. H. Gloor, "Estimation of True Fiber Modulus from Instron Data." AFML Report TM-MAN-68-15, June 1968.
6. R. L. Moore, "An Accurate Strain Measurement Technique for Modulus Determination," Lawrence Livermore Laboratory Chemistry Dept. Note, CDTN 7239, Aug. 1972.
7. T. T. Chiao and R. L. Moore, "Strain Rate Effect on the Ultimate Tensile Stress of Fiber/Epoxy Strands," *J. Composite Materials*, Vol. 5 (1971), p. 127.

Acoustic Emission Produced During Burst Tests of Filament-Wound Bottles*

MARVIN A. HAMSTAD AND T. T. CHIAO

University of California
Lawrence Livermore Laboratory
Livermore, California 94550

(Received May 12, 1973)

ABSTRACT

Acoustic emission was recorded during burst tests of filament-wound, composite pressure vessels. Organic and graphite fibers were tested, and two different epoxy resin systems were used: one with a low and another with a relatively high cure temperature. Acoustic emission was studied for the effects of different winding patterns, artificial flaws, winding-induced fiber fraying, different resins, and different fibers. Small effects produced in the vessels by changes in these variables were greatly magnified when they appeared as changes in acoustic emission. They would, in fact, be difficult or impossible to detect by other test means.

INTRODUCTION

LITTLE PROGRESS HAS been made in the development of nondestructive test methods for filament-wound pressure bottles. Acoustic emission is a promising technique to fulfill this need. But before the capabilities of acoustic emission can be assessed, it will be necessary to gain a more complete understanding of the acoustic emission produced during burst tests of filament-wound pressure bottles. It is toward this interim goal that the work described in this paper was directed. These results are presented with two primary purposes in mind. The first is to point out certain parameters and factors that influence the acoustic emission from filament-wound composites. Second, we hope to demonstrate the insights that can be gained

*This work was performed under the joint auspices of the National Aeronautics and Space Administration–Lewis Research Center–and the Atomic Energy Commission (NASA-AEC Purchase Request C-13980C).

by using acoustic-emission techniques during burst tests of filament-wound composites.

Recently it was reported [1] that for S-glass/epoxy filament-wound pressure bottles, standard acoustic-emission data is sensitive both to changes in the ratio of longitudinal wraps to hoop wraps and to changes in the order of wrapping. The author also concluded that acoustic-emission data could be used to detect a grossly flawed S-glass pressure bottle at about one-third its normal failure pressure. He further indicated that for the S-glass/epoxy system used in those tests, all or nearly all of the acoustic emission (at approximately 70 dB gain) was associated with fiber fracture. This latter result was based in part on the fact that a fracture-toughness, epoxy double-cantilever beam (*DCB*), strengthened with approximately 50 strands of single-end S-glass situated perpendicular to the cleavage plane, produced some 13,000 to 15,000 counts during testing to failure, while an epoxy *DCB* without strands produced approximately 400–600 counts during testing to failure.

SPECIMENS

Composite pressure bottles were wound using two basic wraps. Figure 1 shows drawing of a typical bottle. The first basic wrap, called the longitudinal wrap, was

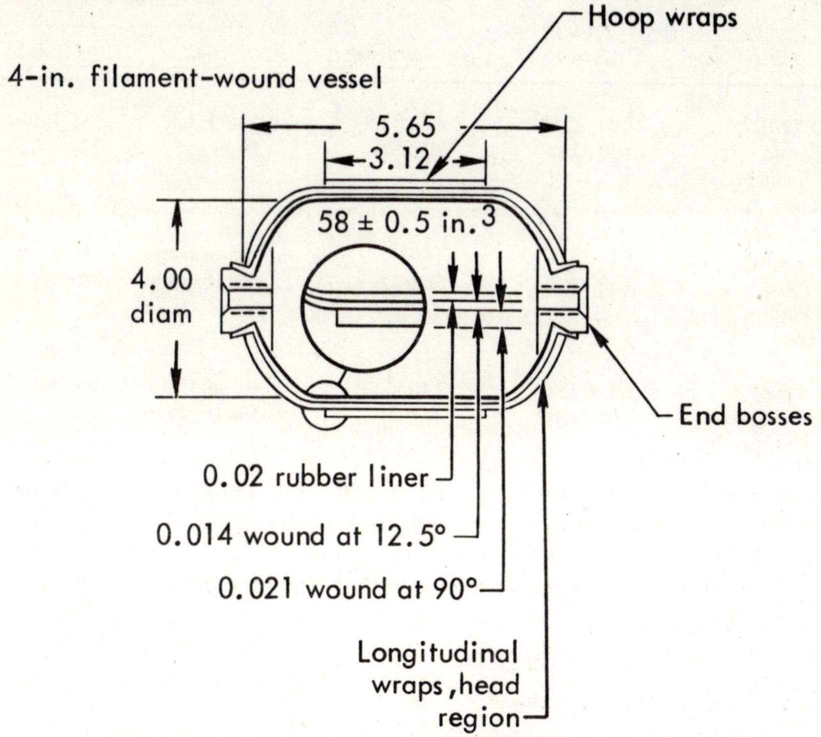

Figure 1. Typical dimensions (in inches) of a filament-wound bottle.

laid down at a fixed angle of between 7° to 13° from the longitudinal axis of the bottle. The second basic wrap, called the hoop wrap, was laid down at an angle of 90° to the longitudinal axis. The hoop wraps only appear on the cylindrical section of the bottle.

One of the primary winding variables is the ratio of longitudinal fiber stress σ_L to hoop fiber stress σ_H. This ratio, which we call the fiber stress ratio, was determined from a simple netting analysis applied to the cylindrical portion of the bottle; it gave an approximate measure of the ratio of the fiber stresses in the two different wraps at a given pressure. Hence, a bottle wound with a balanced design (based on the netting analysis) would have a fiber stress ratio $\sigma_L/\sigma_H = 1.0$. More detailed information on the winding conditions and strength characteristics of the bottles can be found in the papers by Chiao et al [2, 3, 4].

Four material systems were used to fabricate the bottles. These material systems, as well as the cure cycles, are summarized in Table 1.

Table 1. Material Systems

Fiber	Epoxy Resin	Hardener	Curing Cycle: Gell	Curing Cycle: Cure
PRD 49-III[a]: single end, 285 filaments per end	ERL 2258[b] 100 parts	ZZLB 0820[b] 30 parts	140°F for 16 hr	325°F for 2 hr
PRD 49-III[a]: single end, 285 filaments per end	DER 332[c] 100 parts	T403[d] 36 parts	Room temp. for 16–24 hr	165°F for 3 hr
Thornel[b] 400: single end, 1000 filaments per end	DER 332[c] 100 parts	T403[d] 36 parts	Room temp. for 16–24 hr	165°F for 3 hr
Thornel[b] 400: single end, 1000 filaments per end	ERL 2258 100 parts	ZZLB 0820 30 parts	140°F for 16 hr	325°F for 2 hr

[a]DuPont, Wilmington, Delaware

[b]Union Carbide, New York, New York

[c]Dow Chemical Co., Midland, Michigan

[d]Jefferson Chemical Co., Houston, Texas

TESTING AND ACOUSTIC EMISSION SYSTEMS

All the composite bottles were pressurized with fluid. A closed-loop system was used to control the rate of pressurization. All bottles were pressurized at a rate of 1000 psig/min.

A commercial piezoelectric acoustic-emission transducer was coupled to the cylindrical portion of each bottle with a viscous resin. In a few cases a transducer was also coupled to the "head" region of the bottles. The acoustic signal was processed through a preamplifier, a filter (usually set to pass 100–300 kHz), and a power amplifier. In total, the system provided some 55–75 dB of electronic gain as needed. The acoustic signal was presented visually in two ways. First, the signal was displayed on an oscilloscope as a function of time. In addition, a count was made of the number of times that the amplitude of the acoustic signal exceeded a voltage bias (nominally set at 1 V). This count was recorded on an *x-y* plotter as a function of pressure, both as a count rate (usually averaged over 1 or 2 sec) and as a cumulative total of counts.

ACOUSTIC EMISSION TESTS

PRD-49-III Fiber/Epoxy Vessels

Figures 2 and 3 illustrate the two broad types of acoustic-emission records of count rate and summation of counts as a function of internal pressure that can be

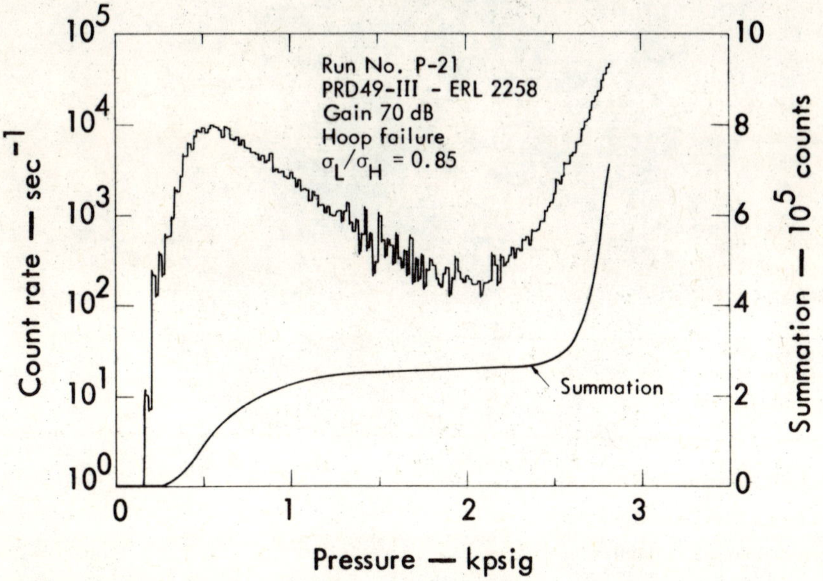

Figure 2. Acoustic emission from an organic-fiber bottle that experiences hoop failure.

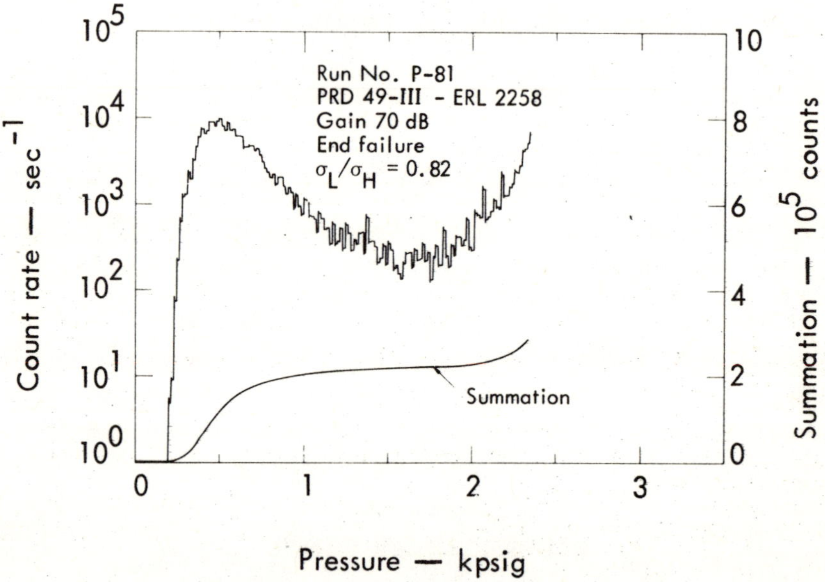

Figure 3. Acoustic emission from an organic-fiber bottle that experiences end failure.

Figure 4. Typical hoop failure of an organic-fiber-reinforced pressure bottle.

obtained from testing of organic fiber composite bottles. The typical predominate failure modes that are associated with these two types of acoustic-emission responses are the so-called hoop failure and end failure. These are shown in Figures 4 and 5. The difference in acoustic emission and failure behavior between these two bottles is associated with how the bottle was wound. The behavior shown in Figures 3 and 5 results from one of the following: (a) a large stress ratio σ_L/σ_H (unbalanced design), (b) a weakness in the fitting region, or (c) failure to reinforce the ends of the hoop wraps. The behavior shown in Figures 2 and 4 results from a sufficiently low stress ratio σ_L/σ_H, adequately reinforced hoop wraps, and no weakness in the fitting region. The typical geometric patterns of acoustic emission shown in Figures 2 and 3 did not significantly change when the wrapping pattern was changed to the so-called interspersed pattern. The interspersed bottle was wound by the following wrapping sequence: two hoop layers, two longitudinal layers, two hoop layers, two longitudinal layers, and, finally, two hoop layers. The noninterspersed bottle was wound by first laying down four longitudinal layers and then six hoop layers. Unless otherwise stated, all bottles were wound with the noninterspersed pattern or with an intermediate pattern consisting of the sequence: two hoop layers, four longitudinal layers, and, finally, four hoop layers.

Figure 5. Typical end failure of an organic-fiber-reinforced pressure bottle.

Figure 6 shows the differences in the summation of acoustic emission versus internal pressure between unflawed bottles that typically experienced end failures

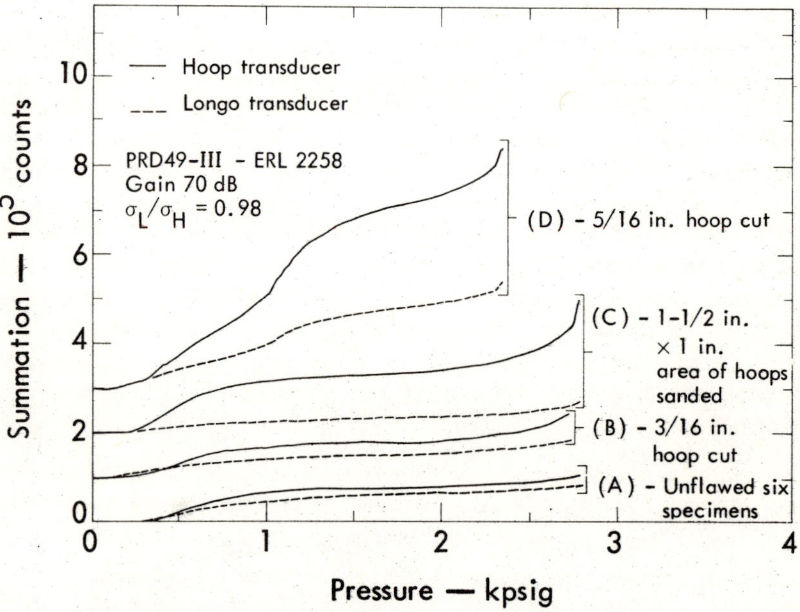

Figure 6. Acoustic emission from flawed organic-fiber bottles.

and bottles with various indicated flaws. In these tests two acoustic emission transducers were used. The transducer on the cylindrical portion of the bottle was called the hoop transducer, and the one on the "head" portion was called the longo transducer. The curves labeled (*A*) represent the typical summation of counts from both transducers for unflawed specimens. Curves (*B*), (*C*), and (*D*) show the typical emission from specimens with flaws: a 3/16 in. cut in the outer hoop wraps (*B*), a 1½ X 1 in. region of the outer hoop wraps sanded with emery paper (*C*), and a 5/16 in. cut in the outer hoop wraps (*D*). For clarity of presentation the zero summation line has been moved up one major division for curves (*B*), (*C*), and (*D*). In case (*B*) the flaw did not control the failure of the bottle. This bottle failed in the end or "head" region, as did the unflawed specimens. The failure location in the other two specimens, curves (*C*) and (*D*), was controlled by the artificial flaws.

Figure 7 compares the acoustic emission from two specimens that differ significantly only in the type of epoxy resin used. In one case the resin was the ERL 2258 system, while in the other case it was the lower-cure-temperature DER 332 system. Some of the properties of these two resin systems, when cured according to the curing cycles shown in Table 1, are summarized in Table 2.

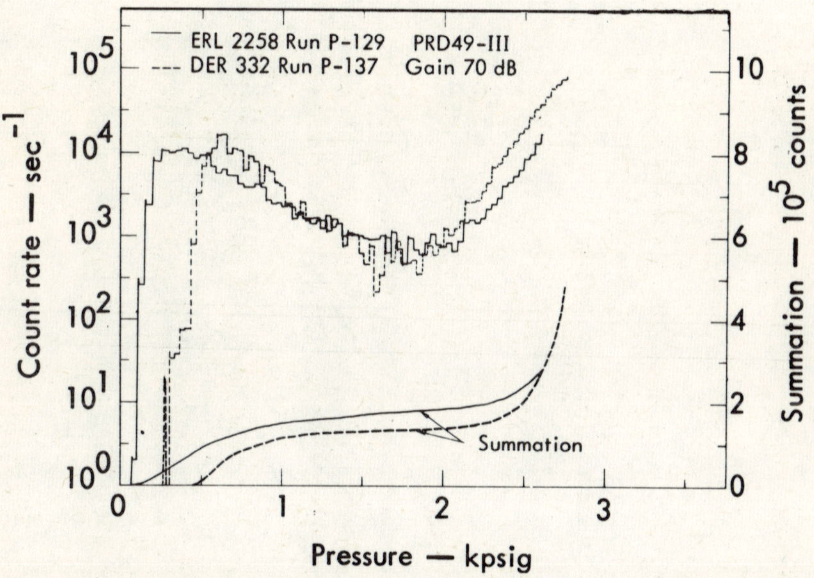

Figure 7. Acoustic emission from organic-fiber bottles with different epoxies.

Table 2. Some Properties of ERL 2258 and DER 332

Epoxy System	*Ultimate Tensile Strength (ksi)*	*Tensile Modulus (ksi)*	*Rupture Elongation (%)*
ERL 2258/ZZL 0820	16.5	557	6.2
DER 332/T-403	10.9	494	5.0

Thornel 400 – Fiber/Epoxy Vessels

Three typical acoustic-emission responses showing summation of counts versus internal pressure are shown in Figure 8 for filament-wound graphite bottles. These three curves represent the typical random variation between supposedly identical bottles. There is no significant deviation from the geometry of these curves when the ratio of hoop wraps to longitudinal wraps (σ_L/σ_H) is changed or when the bottle is wound using the interspersed pattern. The failure mode for the graphite bottles varies in such a fashion that even for identical specimens it is very difficult to find as many as two bottles that have failed in an identical fashion. A bottle that has been tested to failure is shown in Figure 9. Interspersed bottles or bottles with different stress ratios do not have characteristic failure modes either.

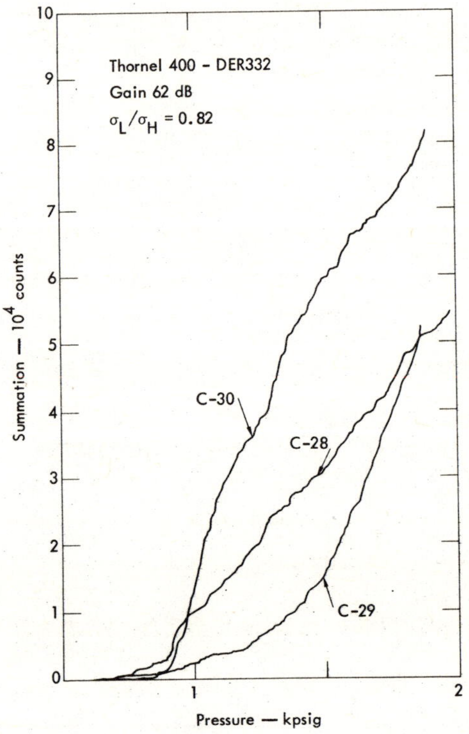

Figure 8. Acoustic emission from graphite-fiber bottles.

Figure 10 compares the count rate and summation of counts for a bottle wound with graphite yarn that tended to "fray" during the winding process with the count rate and summation of counts for a bottle wound with little or no "fraying" during winding.

The effect of the two different epoxy systems on the acoustic emission of otherwise nearly identical graphite bottles is shown in Figure 11. One bottle was wound using the ERL 2258 resin system, while the other was wound using the lower-cure-temperature DER 332 formulation.

DISCUSSION

The discussion and the interpretation of the above experimental results are based on the conclusion that the large majority of the acoustic-emission data gathered at the gain levels used in these tests comes from, or is associated with, fiber failure. The basis for this conclusion was alluded too briefly in the introduction and is discussed more completely in Reference [1]. Although these discussions primarily refer to S-glass-reinforced epoxies, the basic significant finding carries through to organic- and graphite-reinforced epoxies. Namely, the fiber failure stresses are much

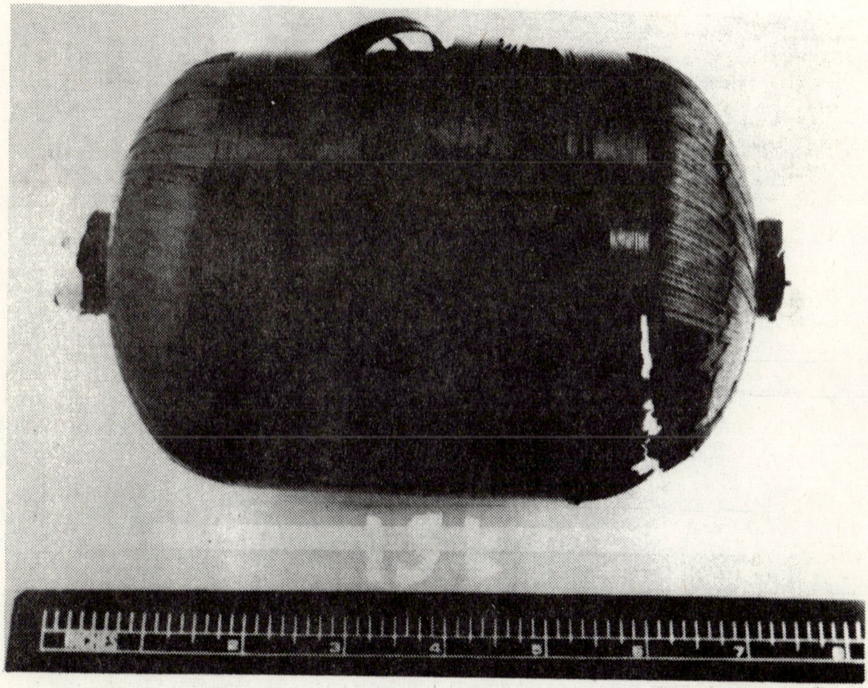

Figure 9. A failed graphite-fiber-reinforced bottle.

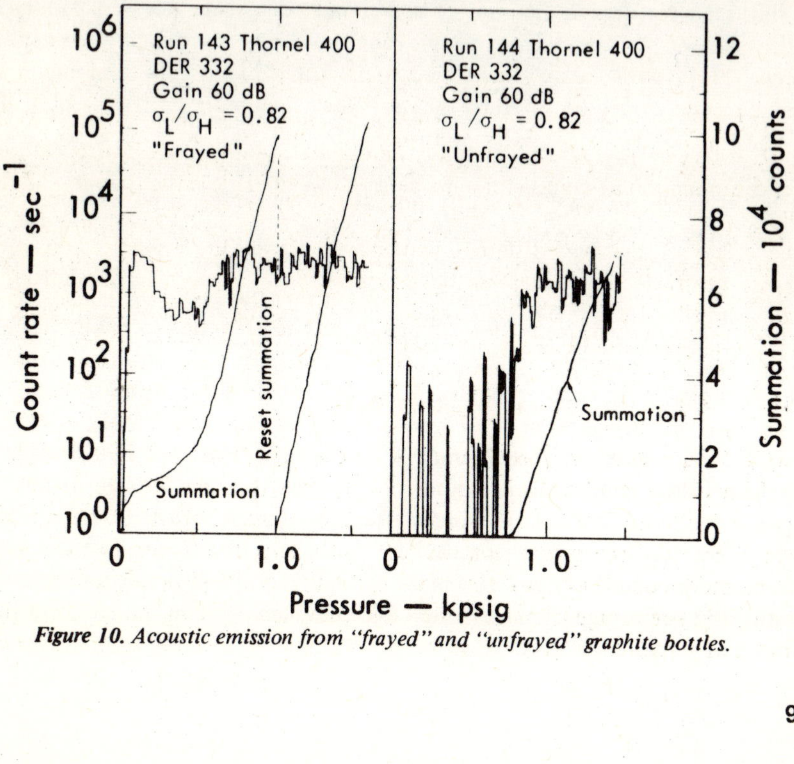

Figure 10. Acoustic emission from "frayed" and "unfrayed" graphite bottles.

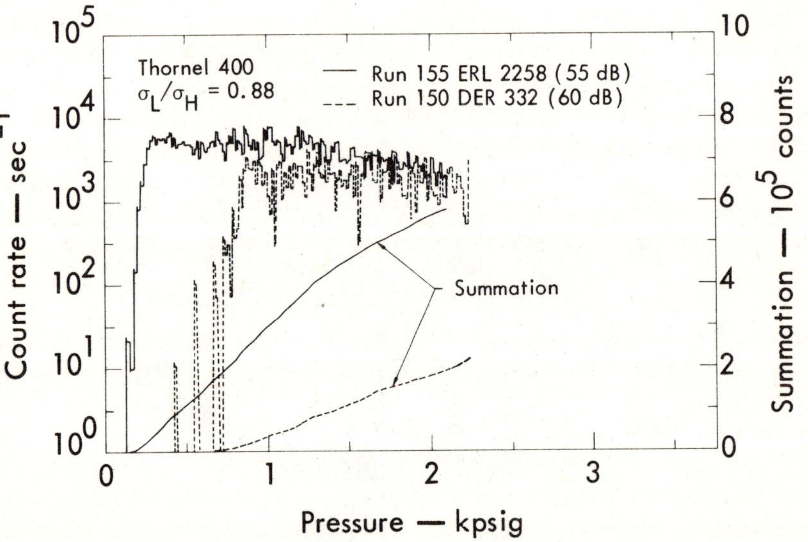

Figure 11. Acoustic emission from graphite fiber bottles with different epoxies.

higher than matrix failure or interfacial debonding stresses, and hence the energy released by these latter mechanisms is much less than that released by fiber failure.

The contrast between summation of counts and count rate shown in Figures 2 and 3 illustrates for organic fiber-reinforced epoxies one of the key applications of acoustic emission to composite pressure bottle design. As was pointed out in Reference [1] (for S-glass-reinforced epoxies), the rapid rise in summation of counts and count rate just prior to failure (shown in Figure 2) results from the many fiber breaks that occur in the relatively uniformly stressed hoop wraps. Thus, the acoustic-emission data can be used for organic and S-glass bottles to distinguish between bottle designs that efficiently load the hoop wraps prior to failure and those that do not. The acoustic-emission behavior shown in Figure 3 indicates that failure in this bottle occurred before the hoop wraps were highly loaded. The three basic causes for this behavior were pointed out in a previous section.

The rapid rise in count rate and summation counts just prior to failure has not been obtained for the graphite fiber-reinforced bottles in spite of several winding pattern changes. This result, coupled with the seemingly brittle failures that typically occur in these bottles (see Figure 9), leads us to conclude that the average number of fiber breaks per unit volume (i.e., the average density of fiber breaks) at which the graphite bottle fails is considerably smaller than the average density of fiber breaks for S-glass and organic bottles. Thus, the graphite bottle fails before the density of breaks undergoes the rapid rise that generates the rapid rise in the acountic-emission data just prior to failure of organic and S-glass bottles. Based on this result, the percentage of single-strand fiber strength that can be obtained in a composite graphite pressure bottle[1] is expected to be lower than that for the S-glass and organic fiber bottles. This observation implies that it may be possible to use acoustic-emission data to determine which fiber-matrix systems are more sensitive (in the sense that they experience complete failure at a lower average density of broken fibers) to the random fiber fractures that occur during loading.

The fact that the basic geometry of the acoustic-emission patterns for the organic- and the graphite-fiber bottles was not significantly altered when the interspersed winding sequence was used is an important observation. This result implies that for these two fibers the interspersed pattern will probably not degrade the performance of such vessels significantly, and may even offer a slight improvement. It was pointed out in Reference [1] that the use of the interspersed pattern in S-glass led to a significantly different acoustic-emission pattern. This pattern indicated a degraded bottle and, indeed, the experimental failure pressures were degraded by some 14–18% below those found for bottles having the non-interspersed pattern.

Both Figure 7 for organic-fiber bottles and Figure 11 for graphite-fiber bottles show that the use of a lower-cure-temperature epoxy system led to an increase in the internal pressure at which a significant amount of acoustic emission begins (defined as the pressure where the summation of counts rises above the zero line). The increase is on the order of two to four times the pressure at which acoustic emission begins in the high-cure-temperature resin bottle. At this point it is only possible to speculate on the cause for this change in acoustic-emission response. It may be that this phenomenon is related to the residual stresses or the fiber matrix configurations that are induced during the cure cycle. In the future a statistical study is planned to determine if the different matrix systems significantly change the average failure pressures of the filament-wound pressure bottles.

The result shown in Figure 10 demonstrates that acoustic emission could be used to check production processes if one desired to sort out graphite-reinforced bottles that had experienced significant "fraying" during winding. It would be necessary to pressurize each bottle to a pressure of only 400 or 500 psig to make this determination. The bottles that show the early acoustic emission could then be separated out. The actual mechanism through which the frayed yarn leads to early emission is not clear.

The results shown in Figure 6 for the flawed organic-fiber bottles indicate that it is possible to detect a flawed bottle when the flaw is controlling the bottle failure. Such flaws may be detected at approximately one-half the normal failure pressure. Note that in test (*B*), where the flaw did not control the bottle failure, that the summation of acoustic emission gave very little indication that the flaw was present. It is also clear from test (*C*) that the location of the acoustic-emission transducer with respect to the flaw location is important. In this case the longo transducer gave no visible indication that the flaw was present. In case (*D*) both

[1] **Based on a large statistical sample of strands and bottles.**

transducers gave adequate indication of the presence of the flaw. The reason for this difference can be partly explained by a difference in the mechanism of propagation of the flaws in cases (*C*) and (*D*). When the acoustic-emission signals from the hoop transducer in these two cases were viewed on an oscilloscope, the indication was that the flaw growth was proceeding in a different manner in each case. The acoustic-emission signals had very large amplitudes in case (*D*). Often the signals completely saturated the electronic equipment. This observation indicated that the flaw growth was proceeding by events involving large energy releases. In case (*C*) the acoustic-emission signals tended to be comparatively small (indicating growth by small energy releases). These rather small signals, which originated in the hoop wraps for case (*C*), were attenuated during propagation to the longo transducer and hence did not have enough amplitude to be counted by the longo acoustic-emission channel. One further point should be made with regard to these tests. In spite of relatively gross flaws, in case (*B*) the flaw did not control the failure, in case (*C*) the bottle failure pressure was not significantly degraded, and in case (*D*) the failure pressure was degraded only some 15%. Hence, it would seem that the organic-fiber bottle is relatively resistant to artificial flaws.

CONCLUSIONS

For S-glass and organic filament-wound pressure bottles, acoustic emission can be used to determine if the hoop wraps are highly loaded before bottle failure occurs.

Acoustic-emission techniques may offer a means to assess the average density of fiber breaks at bottle failure for different combinations of fibers and epoxy matrixes.

Acoustic emission indicates that bottles wound with Thornel-400 may fail at a much lower average density of fiber breaks than S-glass and PRD-49-III bottles.

Acoustic emission could be used to sort out graphite bottles that were frayed during the winding process.

Acoustic emission can sort out flawed organic fiber bottles if the flaw controls the bottle failure.

Use of a lower-cure-temperature epoxy matrix increased by a factor of between two and four the pressure at which graphite- and organic-fiber bottles first produce significant acoustic emission.

REFERENCES

1. M. A. Hamstad, "Acoustic Emission from Filament-Wound Pressure Bottles," Proc. 4th SAMPE Natl. Tech. Conf., Palo Alto, California (October, 1972).
2. T. T. Chiao and A. D. Commins, "Fiber Strength of S-Glass/Epoxy Composites Under Bi-Axial Loading," Proc. 4th SAMPE Natl. Tech. Conf., Palo Alto, California (October, 1972).
3. T. T. Chiao and M. A. Marcon, "Filament-Wound Vessel from An Organic Fiber/Epoxy System," Proc. 28th Annual Conf. on Reinforced Plastics/Composites Institute, SPI., Washington, D.C., February 6–10, 1973.
4. T. T. Chiao, R. L. Moore, and C. M. Walkup, "Graphite Epoxy Composites," submitted to SAMPE Quarterly.